HIPPOCRATES REVISITED

A Collection of Personal Student Oaths

Edited by

Jo Marie Reilly, M.D.

Helena Yu, B.A.

Allan S. Lichtman, M.D.

Rosemary R. Lichtman, Ph.D.

First published in 2015
University of California Medical Humanities Press
in cooperation with The Virtuoso Press

© 2015 by Jo Marie Reilly, M.D., Helena Yu, B.A., Allan Lichtman, M.D., Rosemary Lichtman, Ph.D.

ISBN: 978-0-9889865-8-9 Paperback

Printed in USA

Preface

For many years I have spoken with physicians at all levels of training about the essence of their calling to medicine: what drew them to the profession, what keeps them committed and what helps them stay healthy and grounded in a very demanding career. It is something I have personally reflected on in my journey as a family physician.

I have experienced the power, wonder, gratitude and tremendous compassion that exceptional medical care can bring to patients and their families in both joyous and tragic times. I believe that physicians, at their very core, are motivated by a deep sense of service and dedication to care for others.

Ten years into my own medical training and career, the long hours, demanding call schedule and working with a patient population with few resources became very challenging to balance with my growing family needs and demands. In the midst of my own professional soul searching, I wrote a personalized Hippocratic oath—a testimony to my personal journey and tenets as a physician. This process reaffirmed my commitment to serve my patients and provided an opportunity to reassess my life balance and well being, increasing my awareness of my role as a physician healer.

Shortly after my own medical soul searching, I had the opportunity to apply my experience to medical students. Teaching in the Keck School of Medicine of USC's Professionalism and the Practice of Medicine course, I shared my oath with my first year students as they began their journey, encouraging them to make their own personal commitment to their medical career. Several students wrote their own oaths and shared them with me. They told me what an impression this experience had made on them and encouraged me to formalize the writing project with all the medical students. Directing the course at the time and with the endorsement of the faculty, we initiated a personal Hippocratic oath writing assignment for all medical students. Seven years later, it remains a cornerstone reflective experience for our Keck first year medical students as they begin their own professional journeys. It offers a deeper consideration of the unique path they will chart for themselves as medical professionals.

Finally, I believe the oaths teach us that the relationships our physician healers aspire to have with their patients, in concert with the integrity and sincerity with which they embark on their careers, are powerful forces for healing. This is where the art of medicine begins. It is truly in the humanity of medicine that we are transformed from simply being physicians who cure to ones who restore health. Thus, we become true physician healers.

Jo Marie Reilly, M.D.
Associate Professor of Clinical Family Medicine
Keck School of Medicine of USC

Retiring from our private practices in Obstetrics & Gynecology and Health Psychology, respectively, we were committed to sharing our professional experiences and insight with the next generation of caregivers. As a husband and wife team, we were given the unique opportunity of serving as co-mentors for beginning medical students in the Professionalism in the Practice of Medicine course at the Keck School of Medicine of USC, and it has been our pleasure to teach groups of students for the past six years. The experience has allowed us to interact with talented young students and given us hope for the future.

Reading our students' own versions of the esteemed Hippocratic Oath over the past years, we have been impressed by their remarkable creativity and vision. We felt they were so inspiring that we wanted to collect and publish the oaths so that others could recognize the dedication of our caregivers-in-training.

We shared the idea with the Associate Dean for Curriculum, Pamela Schaff, M.D., who expressed her support. The next day, she met with an interested medical student, Helena Yu, and soon we all joined together with Jo Marie Reilly, M.D. to form the editorial board for this anthology.

We hope that you enjoy the collection and will be as moved as we all were in compiling it. We believe it may even motivate students of a future generation to determine that studying medicine will be their chosen professional path.

Allan S. Lichtman, M.D.
Clinical Professor of Obstetrics and Gynecology
Keck School of Medicine of USC

Rosemary R. Lichtman, Ph.D.
Adjunct Clinical Assistant Professor of Family Medicine
Keck School of Medicine of the USC

"Remember who you wanted to be."

These words are posted in the halls of my medical school, the Keck School of Medicine of USC. I am currently in my second year, and this saying has been especially meaningful. As I reflect on my journey to medical school, I think of how I have always felt a calling to the field of medicine. My motivations lie in the ability of medicine to combine science and art as well as technical knowledge and human interaction. In particular, I am compelled by the power and utility of narrative medicine to transform physicians into more caring, empathetic, and ethical caregivers. Narrative medicine provides a venue for regular reflection and discussion of why decisions were made, what our values and principles are, and how viewpoints differ.

During the first weeks of medical school, I was pleased to have the opportunity to write my own Hippocratic oath in the Professionalism in the Practice of Medicine course. I wrote a letter to myself, hoping to encapsulate my personal beliefs and goals. At that point, I was completely new to medicine, eager to take in every experience. As I move through medical school, residency and beyond, I hope to look back on this letter from time to time and bring myself back to the mindset I was in when I wrote it and remember who I wanted to be.

Serendipitously, I have been able to further pursue my interests in the medical humanities by joining the editorial board of this anthology. I took on this project, hoping to use narrative medicine to promote mindfulness, optimism and ethics at my school. As I look at my classmates, I can see that many of us have been on a long path to decide on a career in medicine, and we are finally following our dreams of making a positive impact on the lives of others. The oaths contained with this anthology certainly reflect these ideals. Each student contributes a unique perspective to Keck, and it is my belief that this anthology will allow us to share our humanity, values, and stories with each other.

Helena Yu, B.A.
Keck School of Medicine of USC
Class of 2017

Introduction

What commitment did your doctor make when first putting on *a white coat*?

Each fall, first year medical students traditionally participate in the White Coat Ceremony. They receive a white coat and recite the Hippocratic Oath, a symbol and a promise respectively of their unique pursuit of medical training. This ritual also serves to highlight their acceptance of the special personal responsibilities bestowed upon them.

What are the basic tenets that young physicians bring to their early careers? Would the process of reflecting on these principles bring a deeper understanding of their calling and a greater recognition of purpose? Is this awareness protective to their soul and spirit as the ensuing years of sleep deprivation, tireless hours of study, discipline and delayed gratification slowly and insidiously challenge them and their commitment?

We believe some of the answers lie in the individual oaths collected in this anthology.

The idea of *Hippocrates Revisited: A Collection of Personal Student Oaths* was inspired by the belief that sharing the hope and vision of young student physicians personalizes the journey of medical professionals. The anthology reflects the diversity and narrative that all student physicians bring to the face of healing and transforming patients' lives. It tells the story of their hopes and dreams as they embark on a professional career.

Sharing the collective spirit and soul of these healers can inspire patients, families and society. We learn from them and one another as we come together to honor these oaths, celebrating the power that comes from integrating compassion into patient care.

A Calling

Most professional life journeys begin with an internal calling, an inner voice governing the choices we make. Why do I find this road so compelling? What draws me to this career path? For physicians, there is a deep passion: the commitment to bring health and an improved quality of life to those whom they serve.

From the early years of medical school through rigorous residency training programs, physicians are molded and shaped by their education, health team, patients and colleagues. Amidst efforts to master scientific knowledge and provide compassionate care, what sustains their continued dedication to medicine and service? What allows them to remain healthy and well while also supporting patients and their families? Physicians must constantly remind themselves of their inner motivations and values. Personal reflection can help re-focus and remind students, residents, fellows and seasoned physicians of their commitment to both patients and self-care.

Physicians who are the most complete, centered and satisfied with their professional lives are those who have a clear understanding of what first brought them to medicine, allowing them to stay rooted in these values. Being mindful of one's calling and abiding by those internal motivations bring tremendous strength to a physician's grounding and practice. It reaffirms the value in the work that he or she does, strengthening the resolve to be true to self and to live by one's code and creed.

Identity Transformation

First year medical students are challenged by their new identity when they enter the hospital setting wearing their white coats. They are often seen as medical professionals by patients who may grant them the rights and privileges of physicians. These expectations can be challenging, as students recognize the dramatic change in how others perceive them. They have entered medical school committed to help patients and realize full well that they do not yet have the ability to provide the needed professional services.

Creating their own oath offers first year students the opportunity to individually come to terms with this inner struggle and professional transformation early on in their careers. They grasp the importance of reflecting on the fundamental question: *Who am I as an emerging doctor and who do I hope to become?* As a first step in forming their particular answer, the students' personal oaths generate a framework to guide them as they build a fundamental set of vows and principles to define themselves and their respect for patient integrity. As students continue to forge their new identity as *physician*, they balance core beliefs and values with the transformative experiences of medical school. These together form a rich fabric they will continue to weave throughout their training and professional career.

Historical Context

Transcending time and cultures, physicians have embraced ethical codes that guide their practice. The most famous documents that set forth standards governing physicians' ethical and moral practices are the Hippocratic Oath (Appendix 1) and the Oath of Maimonides (Appendix 2). The first, the Hippocratic Oath, dates back to between the 7th century BC and the 1st century AD.[1] Hippocrates, revered as the father of western medicine, was a philosopher and naturalist as well as physician. The original version of his oath is divided into two parts: (1) the duties of the medical student towards his/her teacher and that teacher's family and (2) the duties, morals and ethics of a physician in treating his/her patient. While the early Greek physicians largely ignored the Oath, it was accepted by the Judeo-Christian culture and has served as a moral compass for physician practices through the years. The Oath held a physician to a just and ethical medical practice.

The Oath of Maimonides[2] was written toward the end of the 12th century by Rabbi Moshe ben Maimon, known as Moses Maimonides. A physician and scholar of Jewish law, he was the most important Jewish philosopher of the Middle Ages. Maimonides' oath reveals his belief that the practice of medicine is a sacred calling, with physicians' abilities granted by God to aid their patients. He also addresses the important intellectual challenges of studying the science of medicine as well as the art.

Another treatise, which dates from the 9th century, is foundational to early medical professionalism. This Hebrew document, "The Book of Admonitions to the Physician"[3] is attributed to the Judeo-Arabic physician Isaac Israeli. This series of pragmatic and idealistic admonitions, among others, requires a physician to obtain continuing medical education, pursue excellence in care delivery, maintain self-care, and practice the art of medicine.

In China, Sun Simiao, known as "the king of medicine," completed *On the Absolute Sincerity of Great Physicians* during the Tang Dynasty, instructing that physicians treat all patients equally.[4] The ancient vow taken by Hindu physicians, Vaidya's oath, mandates that they put the needs of their patients before their own.[5] Similarly, Buddhists take a physician's oath, known as the Vejjavatapada.[6] Beginning in the 16th century, Japanese physicians adhered to the 17 Rules of Enjuin.[7] Still another oath is espoused by the Muslim tradition.[8]

Fundamental medical ethics are rooted in these early documents, and they set the tone and standards of current physicians' practices. In a world expanding with more technology and scientific advances than those early physicians ever could have imagined, we are still governed by Hippocrates' timeless statement, "(while) the art (of medicine) is getting longer and longer, the brain of the student is not bigger and bigger."[9] These early medical creeds, "sought to give society at least some guarantees…and assure it access: to competent medical care, to humane treatment, to preservation of personal dignity, and to maintenance of strict confidentiality."[10]

Ethics in Current Medical Education

While the Hippocratic Oath has been modified over the centuries, it has remained an integral element of medical ethics. By 1993, 98% of U.S. medical schools administered a revised Oath, a significant increase from 26% in 1928.[11] It is currently recited at medical schools across the country as their students take part in the historical rite of passage, the White Coat Ceremony. Recitation of the Keck School of Medicine's adapted Hippocratic Oath[12] (Appendix 3) begins the students' journey in their first year. They recite the Oath again when they transition from the preclinical to the clinical years and once more at graduation.

Not all of the oaths recited today evoke deities and pledges to higher powers or contain admonitions concerning abortion and euthanasia. Perhaps the newer oaths carry less weight in modern society, with its fewer absolutes and greater moral relativism.[13] Yet the recitation of the Hippocratic Oath allows for a pause and a time of reflection. It offers an opportunity for young medical trainees to explore the ethics, values, and professional duties expected of physicians.[14,15]

We believe this is significant, as professionalism is one of the core competencies emphasized in medical training today.[16,17] Adherence to the tenets and ideals espoused by the Hippocratic Oath is integral to a medical student's training throughout the country. These doctrines offer hope in the changing world of medicine that is threatening to erode the fragile doctor-patient relationship so intrinsic to the essence of the work of physicians. This challenging health care landscape begs for a voice of compassion, reason and simplicity. It cries for inspiration from a new generation of physician leaders who reaffirm the values of patient care and commitment to the profession.

In compiling this anthology of student Hippocratic oaths, we hope to honor the richness of individual narratives and the unique contributions that all medical students bring to their profession.

Hippocratic Oaths Today

At the Keck School of Medicine of USC, students are asked to write their own Hippocratic Oath as part of their professionalism course. These individual oaths offer a deeper reflection on the unique path that they will chart for themselves as medical professionals. Each one represents the student's personal testament to the medical profession, community, self, patients and society. It is an affirmation of who they are at the beginning of their medical studies, and who they hope to become.

To celebrate the serious dedication and creativity of these professionals in training, we have compiled an anthology of original oaths, composed over the past seven years by different classes of medical students during their first month of school. They reflect the tremendous metamorphosis of individual medical students at a pivotal moment in their journey as healers. The oaths highlight the students' perception of how they intend to guide themselves in their professional lives. Some of the oaths are promises, others are declarations, songs, poems, or works of art. Some students elected to be credited while others chose to remain anonymous. All are expressions of how new healers intend to fulfill their exceptional responsibilities to themselves, their families, society, and most especially their patients, as originally espoused by Hippocrates.

These expressions of optimism and promise are deeply encouraging and are likely representative of medical students throughout the country. We believe that this anthology offers inspiration to new physicians, patients and their families. The oaths that follow remind us that physicians are called to be healers. And they answer that call with great humility, dedication and sincerity.

The Healer's Creed

Divine Healer, grant that I may always seek:

To listen completely, so to better allow a patient the opportunity to express their pain, needs and discomforts.

To understand, when compassion and caring encourage and support a patient's simple desire to be heard respectfully and attentively.

To respect differences in faith, gender and ethnicity, when turning a blind eye perpetuates ignorance and mistrust and magnifies the injustices in healthcare delivery.

To advocate for the homeless, the hungry, the despairing, those too broken to see the light and hope shining through their darkness.

To stand for justice and truth, when blinded by the politics and temptation to choose the path of least resistance.

To touch, when touching communicates presence, understanding and compassion.

To empower the victim to seek courage and support, when paralyzed by fear and immobilized by violence.

To find patience and mercy, when challenged by the anger of a difficult patient, the patient plagued by a personality disorder or the frightened addict embracing his/her denial.

To treat the addict, the survivor, the prisoner, the elected official all with the same dignity and respect, merely because they are all worthy.

To redirect the anger of depression, instead to fight for self-preservation, self-discovery and self-love.

To have the strength to encourage and motivate unhealthy patient behaviors and lifestyle choices, offering healthy choices and evoking a patient's spirit and will to change.

To communicate honestly, conscientiously and in a language that patients can comprehend, when failure to do so exacerbates both patient's noncompliance and fear of illness.

To place significant importance on self care, both valuing the need to be a role model of good health and prioritizing personal wellness.

To have the courage to confront our broken colleagues confidentially and compassionately, when to neglect this duty undermines our oath as healers and jeopardizes the well being of others.

To teach our students and residents not only the pathology of disease, but the art of medicine and the joys of our profession.

To rediscover the miracle of life at the birth of a child and in a parent's delight and discovery of their newborn.

To relive the world through the innocence, curiosity and simple pleasures of a child.

To encourage the troubled teen, struggling for independence and challenging limits while simultaneously begging for direction and caring.

To hold the hand of a woman in labor, remembering that alleviating suffering and sharing kindness are often most meaningful in life's small tasks.

To help the elderly age gracefully, respecting their wisdom, sharing their stories and accompanying them as their bodies age.

To palliate, when to radiate forgets the dignity of the soul, the essence of the human spirit.

To heal, when curing denies the hearts cry for the simple comforts of love and caring.

To embrace our gifts and weakness as healers so as to better understand human kind.

To minister with virtue, integrity, self-awareness and wholeness.

Divine Comforter,
Guide us on our walk with patients.
May we journey with invitation and acceptance,
Knowing that the art of medicine is truly a divine gift
A grace to be shared and treasured. A Blessing.
A vocation. Life-giving.
And so we embrace our role as healer graciously,
And we are transformed.

Jo Marie Reilly, M.D.
July 2002

Reflections on Living My Hippocratic Oath

I officially began my medical career when I recited the Hippocratic Oath at my medical school graduation. It was both a cherished privilege and the fulfillment of a boyhood dream. Since that moment, I have tried to apply the values the Oath demands by patterning myself after my own caregivers.

I was inspired by my Pediatrician, whom I loved and admired. He treated me, a sickly first-generation immigrant child, with the utmost concern and was clearly dedicated to my wellbeing. He related to my mother and grandmother respectfully, allowing them to diligently care for me.

My parents also illustrated how to value and care for others through their interactions with employees and customers in our family bakery. My father was an especially important role model, voluntarily caring for a special needs homeless person who had been abandoned by his family. Dad found him an apartment, gave him a job and served as his surrogate parent over many years.

These vivid examples of how to value and care for other people guided me to become a healer, unaffected by a patient's social class or economic situation. Given the distinct opportunity to become a physician, I appreciated being able to give back to society as well as honor my parents and the doctor who cared for me. I committed myself to rendering that same respect to each and every patient I had the privilege to serve.

The specialty of Obstetrics and Gynecology afforded me a unique experience in that I was able to become part of my patients' families in very special ways. I enjoyed the continuity and experience of helping to deliver a baby, later take care of her as a teenager, and care for her mother and grandmother during all those years. I was honored to be a part of the joys and sorrows of my patients.

My personal oath also included a responsibility to teach. Retiring from my solo private practice after 30 years, I felt I needed to share some of the knowledge and experience I had accrued by teaching residents and medical students. I had been appointed a Clinical Professor at the Keck School of Medicine of USC some years before and when I left private practice, I was able to follow my dream of educating the younger generation about the art and science of medicine.

I've now come full circle, helping these young, enthusiastic physicians-in-training create their own compelling personal oaths.

Allan S. Lichtman, M.D.
2014

STUDENT OATHS

Dear future Helena,

Today, you finished your first two weeks of medical school. Already, you've been bombarded with piles of new information: pyruvate dehydrogenase, diabetes mellitus, CFTR mutations, and the like. You've also been assigned to write your own version of the Hippocratic Oath, which is this letter to yourself.

Even though it seems like medical school only consists of endless lecture hours on basic sciences, it takes more than a mastery of basic sciences to learn how to become a physician, since science is just one part of the puzzle of becoming a physician. You need to learn how to infuse science into the day-to-day human interactions and relationships that you're forming as a physician-in-training. Challenge yourself to uphold all the qualities that are important to you as a person and transfer these to your identity as a physician. Better yet, commit to melding the two into a new identity, since you've chosen an ancient and noble profession that is, as you know it right now, inseparable from your individual identity. Hold yourself to the highest standards of honesty, integrity, and accountability. Stand up for what you believe is right, even if it is unpopular to do so, or the hard stance to take. Never sit idly and silently; vow to fight for people who don't have a voice because, as their future physician, you may be their only advocate.

Promise to listen empathically and give your patients the best version of yourself and your abilities and the most you are able to offer. People come to see you when they're unwell and unhappy, worried and nervous. Work to become more optimistic and positive, and reach for the day when your best side is your most natural inclination, when you won't need to force yourself to remain optimistic and positive, and when this "best" self becomes second nature. Despite the limitations of the health care system, always be hopeful that you can make a difference. Persevere through discomfort; in fact, embrace it. Commit a minimum of 100% of yourself to everything you do and always be willing to learn and try new things. As you've learned many times through your life: if something feels too easy, you probably aren't trying hard enough. Let go of the idea that everything will come easily to you, and brace yourself for challenges. After all, as a future physician, you are in a unique position within the community as, at once, a caregiver, scientist, advocate, teacher, and leader.

Ultimately, you're here to learn how to become a physician. Remember the self you were when you came into medical school, and hold tightly to your optimism. Always bring yourself down to earth when you feel overwhelmed, scared, or lonely. Realize that while you picked a hard journey for yourself, it will, as with everything you do, be what you make of it, so make it worthwhile. Think back to the early days of your medical education and the excitement and idealism you had as you received your first medical school acceptance and as you eventually decided to matriculate at Keck. Always hold tightly to the fact that you chose this profession because you wanted to serve others, contribute to the progress of society and advancement of medical knowledge, and in the end, take care of other people.

Sincerely,
Helena circa 2013

Helena Yu

As I don my white coat and take the oath of Hippocrates

I vow, first and foremost, to make certain guarantees:

To put my patients first and uphold the tradition

Of being just and generous to all my comrades in medicine

To conduct myself always with honor and integrity

And be upstanding and honest to the highest degree

To steadfastly search for a solution or cure

And never do harm or take a questionable detour

To listen closely to my patients and offer advice

Without divulging their secrets, no matter the price

To show them compassion and learn to empathize

So I can advocate for them as one of their trusted allies

To dedicate myself to this remarkable art

And strive to improve if I happen to fall short

These are the promises that I am willing to make

Proclamations that I will never bend or break

Becoming a doctor is the end goal

So I can improve the health of society as a whole

Although this journey might be long and tough

The prospect of saving lives is more than enough

To motivate me to be the best I can be

At the Keck School of Medicine of USC

Swetha Ramachandran

The following represents my values, personal commitments, and important thoughts as viewed from the perspective of words...there was just too much emotion to be limited by prose.

Shayan Rab

I, Emily Levy, on this 14th of August 2011, promise to be exceptional to my patients.

I will remember that medicine is an art as well as a science and I will strive to master my craft. I will connect to my patients, treating the person and not the disease. I intend to consider the psychosocial components of health and incorporate public health into the practice of medicine. I will be kind to my patients. I will maintain my compassion, empathy, optimism and commitment to the underserved. I will establish trust with my patients and colleagues. I will contribute to scientific knowledge through research.

I allow myself to be open to the lessons my patients and peers have to teach. Likewise, I will educate and share my knowledge. I will listen to my patients with nonjudgmental ears and maintain the utmost security and confidentiality with any information they share.

I will express cultural sensitivity, respect, and honesty with my patients. I understand differences in culture and language may be an obstacle to care, but I will work to overcome those obstacles. I will keep an open mind to alternative medicine I do not know and seek out an understanding of these treatment options. I will work with the patient to optimize care. My priority will be the welfare of my patients.

I will live my life as a physician, and person, with integrity and professionalism. I will use my best judgment in all decisions, be just in my actions, and never cause purposeful harm. I will fight for the rights of my patients. I will treat for the sake of patient improvement, and not physician payments. I will come to the aid of both my patients and those people in need whose path I cross in my private life.

I will learn from each patient encounter. I commit to a life of learning and scholarship, and a willingness to seek help and question the uncertain. I will remain committed to personal and professional improvement, reflecting on my actions and choices as a physician. I will not attempt anything beyond my personal capabilities and will reach out to my healthcare team when in need. I vow to take ownership for my actions and remain accountable for any mistake. I will heed all advice. I will be a team player.

I promise to be kind to myself in my pursuit of medicine. I will maintain balance in my life to fuel my emotional health. I vow to find an alternative pathway if I experience burn-out or ever become unhappy in my chosen field of medicine. I will do this for the sake of myself and for my patients. I allow myself to have a personal life, but will maintain a separation between my private life and professional work.

I will know my limitations, acknowledge my strengths and improve upon my weaknesses. I will strive to accept the uncontrollable. I will be emotional…at the right moments.

I promise to be exceptional to my patients.

Emily Levy

I vow to uphold the values of professionalism, insight, honor, and compassion in the practice of Medicine; to pursue the course of action that is in the best interest of my patients; to be an advocate for the voiceless in distress; and to work with my colleagues and my community to address the larger-scale issues that afflict the health of many.

I commit to a life of learning through thick and thin; to mentorship for others and for myself from my patients, my colleagues, my family, and my community; and to always respect and uphold human integrity.

I look forward to hearing my patients' stories; to seeing their smiles; to shaking their hands. I look forward to working with my community to guide lives towards a healthier and happier future. I look forward to learning, discovering, and applying the latest advances in medical knowledge, technology, and research for the betterment of the men and women who come under my care.

Finally, I promise to bring a measure of clarity, comfort, and calm to those who are in distress; to explain to them the situation at hand; and to address the pertinent issues that arise in my patients to the best of my abilities. I promise to be an honorable physician, a caring friend, and a staunch advocate for the health of all.

Such is my oath; may I keep it forever with the same if not greater integrity and vigor of my first days of medical school.

Alice Kim

Humble
I vow to remain humble in my practice of medicine: the intrinsic value of human life is ubiquitous across profession, race, and creed.

Innovative
I intend to help progress the profession through being innovative and embracing the fluid nature of medicine to improve the quality of life of my patients as well as patients of future generations of doctors.

Professional
I will hold myself to the highest professional standard possible.

Patient
I will remain patient in my practice of medicine and understand that successful diagnoses often lie in the minutia of a patient's life that may superficially seem arbitrary.

Open-minded
I will remember to keep an open mind when interacting with patients. There is always room for improvement and growth through the understanding of different perspectives.

Compassionate
I strive to be compassionate towards everyone I interact with, both patients and colleagues.

Responsive
I promise to be responsive to a patient's needs. I will sacrifice my comfort and convenience for those who have placed their trust in me.

Amicable
I will remain amicable towards my patients where I am seen as a health care provider and confidant.

Trustworthy
I will hold myself to being true to the teachings of the profession and the confidentiality with which patients have trusted in me.

Intelligent
I will do nothing to diminish my intelligence by partaking in activities that could jeopardize any of the above mentioned characteristics of a great physician.

Confident
I will remain confident in my professional and moral frameworks, which are inherently intertwined.

Abraham Kaslow

I solemnly promise that I will fulfill my duty as a physician to the absolute best of my given skill and ability.

And while I may be designated as a doctor, it should be by title only, as this is not simply a job or trivial responsibility, but rather a noble calling that requires a complete dedication of my life for the sake of humanity.

I will neither provide harmful treatment, supply poisonous drugs, nor fail to bestow the best quality of care possible. I pledge that at times when I do not know what to do, that I will forgo my ego and ask my companions and colleagues for assistance, simply for the sake of my patient.

I must realize that my ability is mortally limited. I shall not believe in any way that I can "play God" and that the choice of life or death ultimately rests with God.

I will be a caregiver and hope-provider above everything else. Regardless of what may happen, let me never forget that I am at the mercy of those whom I care for.

I vow that regardless of the color or creed of my patient, I will treat each and everyone with kindness and respect fitting for all of God's children.

Moreover, let my patients be my most important teacher. They know their body best, and it would behoove me to listen to them and take to heart their feelings.

Likewise, let me always be completely truthful to the patient and their family. During these times of mistrust of physicians, I have an even greater duty to be morally ethical, even if it may mean that I must admit to a mistake.

It is an honor that society has allowed me to take care of humanity. I keep this oath faithfully and promise to respect this honor to my fullest ability from here on foreword.

I vow to devote myself to truly embodying the idea of leader as servant.

To be worthy of trust - the honest communication of fears, pains, and hopes - and to use a quiet commitment to empower patients to find the strength already within themselves. To provide the healing ministry of presence with empathy. To speak with gentleness and thoughtfulness. To be humble, generous of spirit, and fiercely protective of the hope, innocence, vulnerability, dignity, and humanity of my patients. To shine the light of consciousness on health and provide awareness – the greatest agent for change.

And when in doubt about what to do, to be compassionate and loving. Love - the great healer of scars and the answer to everything. I vow to look for ways to offer it and ways I can be it. To be love... kind words, a blessing, sweetness, prayer, laughter, grace, and gratitude. To guide my patients to give birth again to their dreams, to sing the song of their hearts, and live in the hope of truth, light, physical and spiritual nourishment.

The horizon leans forward, inviting us to place new footprints of change. I vow to forge a path of compassion, always bringing my light, however small, to places where it's dark and leading my patients with passionate enthusiasm to run, dance, and soar greatly into the miraculous heart of life.

Anna Ter-Zakarian

For the Rest of My Life

For the rest of my life, I will not simply be a doctor; I will strive to be the best physician I can be, a healer, a teacher, a friend, a confidant, an equal.

As a pediatrician, I will make the experience of going to the doctor's office a pleasant one, a warm one, sometimes even a fun one. I will be welcoming and full of smiles. I will have a deep understanding of what it's like to be a small child, scared and confused, anxious and in pain. I will make sure the patient knows I'm her friend, that she doesn't need to be afraid, that I only want to help. I will explain what I'm doing to ease that confusion; why I'm checking this little boy's ears or why I must poke this needle in his arm, but just for a second. I will make the experience as painless as possible, yet offer comfort and assurance when pain is unavoidable.

I will ease the mind of the parent, often more confused and more anxious. I will explain in detail my thoughts and ideas, I will carefully explain the options for treatment, I will assure the parent that the precious child will be fine in a few days… or I will offer comfort and compassion when the outlook is dim.

As a teacher, I will make sure no patient sits through a physical examination without knowing what I'm looking for. I promise to allow no patient to leave the office without knowing my diagnosis. I will explain how to take medications correctly and what they are used for. I will teach the importance of good health and taking care of one's self. I will never let a diabetic patient go home without knowing the vast importance of eating a healthy diet and maintaining good blood sugar levels. I will never let a patient go home more confused than when they arrived.

As an equal, I will make sure that any patient who feels undeserving of medical care, who feels inferior to me, or who feels worthless, will leave the appointment with a more optimistic outlook, if only slightly. I feel it is my duty to provide my services to patients in underserved communities, even in other countries. It is my duty to do what I can to try to improve the health of all patients, regardless of race, gender, sexuality, appearance. And I will make it my life goal to provide as much medical care and kinship as I can to those less fortunate whose life experiences have deceivingly taught them that their lives are worth less than the lives of others.

As a patient, I have felt that same sense of uneasiness and dread that my future child patients will feel. As a young child with serious medical conditions, I have seen my parents speak anxiously with numerous doctors, yearning for a way for me to get better. As a student volunteer at clinics and hospitals, I have encountered too many diabetic patients who eat doughnuts for breakfast and too many patients with hypertension who do not know why they take atenolol each day. Worst of all, I have seen and met too many impoverished families who simply accept the idea that their lives are not worth the time of dentists and pediatricians and family doctors and nurses. Whose lives seem to be valued less in society because of their economic status or the tattoos all over their bodies or the color of their skin.

Because of what I have seen and experienced, I have come to realize the kind of physician I want to be one day, and I will continue to learn how to best take care of my patients every single day for the rest of my life. I will strive to serve as a beacon of hope to those whose hope and self-worth has been stripped away from them. I will strive to be the best physician I can be, a healer, a teacher, a friend, a confidant, an equal.

Kelsey Krigstein

I shall begin only this sentence with "I."
From hereon, the patient shall I place before myself.

                                                           Listen to the patient's story;
                                    hear the heart as well as the heartbeat.

Speak the patient's language;
communicate to be understood, not merely obeyed.

                        Know also when not to speak;
              protect what is not yours to tell.

                                      Be attentive, observant, perceptive;
                                            see without making judgments.

Look outwards,
and let your gaze include the patient's lifestyle, future, and family.

                      Then look closely,
           and note always the hands, palms, fingers
                  that are so much like
                          your own.

                                    Take those hands in yours;
                              your hands will not only treat,
                                      but also support.

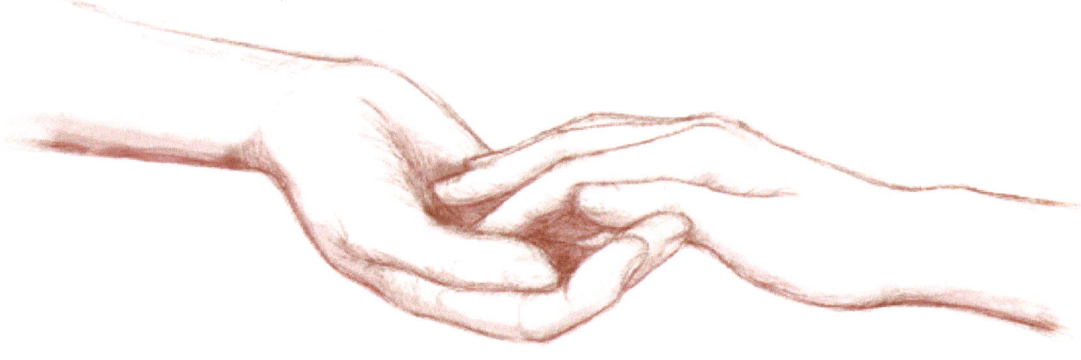

Danielle Li

I do solemnly swear, upon my honor and my conscience, to the best of my ability and judgment, this oath:

I will make the care of my patients my first concern, always acting with care and compassion in their interest.

I will treat every patient politely and considerately, never letting go of basic civility.

I will respect patients' dignity and privacy, ensuring that I am sensitive and discreet in all that I say and do.

I will listen to patients and respect their views, taking their input seriously and working with them.

I will give patients information in a way they can understand, ensuring that they can comprehend what I tell them.

I will respect the rights of patients to be fully involved in decisions about their care, creating a partnership with each of them.

I will prevent disease whenever possible, knowing that prevention is preferable to cure.

I will keep my professional skills and knowledge up to date, practicing evidence-based medicine whenever possible.

I will recognize the limits of my professional experience, doing what is in the best interest of my patients, not my ego, and never being ashamed to ask for help.

I will make sure that my personal beliefs do not prejudice my patients' care, never judging a person by his or her given situation.

I will act quickly to protect patients if I believe that I, or a colleague, may not be fit to practice, while never denigrating a colleague publicly or to patients.

I will avoid abusing my position as a doctor, serving without self-interest and always remembering that I am privileged to help others.

I will work with colleagues, as a part of a team, in the ways that best serve patients' interests.

I will be mindful of the art of medicine, treating patients with warmth, sympathy, and understanding.

I will think of each patient as a person, a human being, rather than a medical case or set of symptoms and problems.

I will be honest and trustworthy, ensuring that my patients feel comfortable entrusting me with their lives and well-being.

I will, at all times, treat the patient as I would wish others to treat me.

I will constantly strive to honor this oath in my service as a physician.

So help me God.

Anonymous

I vow to treat my fellow man with dignity and respect regardless of creed, color, or background

I will practice my craft with precision and purpose

I let go of prejudgments that cloud truth

I promise to be removed from corruption and vice

I value altruism over personal gain

I will use compassion to help patients through hardship

I vow to keep confidentiality and patient integrity at the forefront of my mind

I commit to the empathetic treatment of all

I value honor and respect and will not violate the trust that others have in me

I will not waiver from these ideals and commit to them with utmost conviction

Braden McKnight

MY OWN HIPPOCRATIC OATH

17 April 2008. USC Keck School of Medicine. California. Palm Trees. Hot Weather. Use my gifts to serve others. Life-long learning. Take the time to inspire and be inspired by others. I can learn new things from my peers, professors, and patients.

Patient centered care. Empower my patients to participate in decision-making at every step of their medical care. Listen to their stories. Empathize with their situation. Provide an array of well-informed choices. Honor their objectives in life. Provide patients with the right people and resources. Respect their final decisions. Answer their questions.

15 August 2008. White Coat. People will open up to you. Surround myself with interesting people. Eliminate the words *busy*, *stressed*, and *tired* from my vocabulary. Think beyond conventional borders. Be creative. Be optimistic. Remember that healing the spirit is as important as healing the physical body. Make a world of difference in people's lives. Pray for my patients. Trust in the Lord with all my heart and lean not on my own understanding. In all my ways acknowledge the Lord and He shall direct my paths. Proverbs 3:5-6. Practice with humility, joy, and encouragement. And a sense of humor. Live with integrity even when no one is looking. Remember to say thank you to Mom and Dad for their constant support.

Love is patient. 22 August 2008.

As a future doctor, it's our responsibility

To use the knowledge from med school

In healing patients to the best of one's ability

While maintaining composure and abiding by every rule

Because you see, sometimes patients don't comply

Physicians are under constant pressure

But despite this, there's one thing we can't deny

It's our job and passion to make them better

Practicing medicine is an art form

One must follow the guidelines set forth in HIPAA

While bearing the burdens of health care reform

No matter who we treat, from the needy to the bourgeois

So onto these actions I swear an oath

That I'll fulfill these deeds, even when removing tumorous growths

But all of this aside

I know I'll practice with pride

As medicine is something I could never loathe.

Dennis Guo

As a budding physician, I must take the time now to honor, respect, and reflect on the important role that I shall be taking on. Once I slip on that oh-so-powerful and authoritative white coat, I must always remember the responsibilities that I have not just to my patients and society, but to myself as well. I vow to never lose track of my ideals or values and to never forget the bigger picture of life.

I promise to treat all of my patients as human beings. When they come to me for care, I will look at them as my mother, my sister, my father, my neighbors, my friends and ask myself, "Who am I to act better than anyone because I have a white coat on? Who am I to judge a stranger's life based on one short encounter with them? And who am I to forget the outside world, untouched by the stress and perfectionism that will be my career?" I will not be afraid to touch my patients even if it is just a reassuring pat on the arm or a thoughtful handshake.

I vow to never forget where I come from, regardless of the social class that my career will take me. As a consequence of this, I understand my personal obligation to reach out to my community and to act as a role model for young children who may be unable to visualize a brighter future. I will take on this role with dignity and integrity, realizing that I am always on stage and that my actions will be held to higher standards as I now represent a professional of the medical community.

I shall always remember to take care of myself as well as my patients, for I cannot care for others if I am unhealthy physically or mentally. I hope to always remember that being a doctor is not my life, but rather a part of it and that I deserve to be a diverse and creative human being with interests outside of medicine. May I never forget the beauty of life, the joy in living, and the inevitability of death. And most importantly, may I always KEEP IT REAL.

Miriam Lassiter

Mariana Martinez

As a physician, I promise to treat my patients with utmost respect and compassion; each patient is important, an individual who deserves to receive not only an accurate diagnosis but also the understanding of his or her physician.

As a physician, I promise to dedicate myself completely to determining the correct diagnosis for my patient. This process begins now, as I study and learn; in order to have the most complete set of skills and knowledge I need to commit myself to daily and lifelong learning, beginning in medical school and continuing until the final day I am within the practice of medicine.

As a physician, in making the correct diagnosis and caring for my patient, I promise to call on all of the skills I possess such that said can occur in the most efficient and effective manner. I will be quick-thinking, steady-handed, patient, and poised under the pressures of time and emotion.

As a physician, I promise to work cooperatively with my colleagues such that the most efficient and effective care can be provided. I will both lead and work as part of a team, in whatever setting I practice medicine.

As a physician, I promise to see the bigger picture of medicine. Not only will I strive to care for my individual patients in an accurate and compassionate manner, I will also commit myself to improve the systems in which I practice. I will observe the points in care processes that slow the delivery of care to patients, or that lead to error. I will strive to identify areas for quality improvement, such that I can positively impact more patients' lives, and the larger scope of medicine.

As a physician, I promise to advocate for those who are disadvantaged in our societies and societies around the world. Locally, I will serve those who cannot afford health insurance, or those who come from less fortunate backgrounds. Internationally I will contribute to medical aid in the highest capacity I am able. I aim to keep a global perspective in medicine, and strive to serve those whose access to medical resources is limited.

As a physician, I promise to study and engage myself in the art of medicine. I will use my previous studies and knowledge of fine art to guide me through the concepts of diagnosis. I will maintain the ideal that art is medicine, that some areas are inevitably and perpetually gray, that things may be more abstract than perceived, that a picture can shift with varying vantage points, that there are aspects of clinical medicine which transcend the realm of science, and occasionally that of human experience.

As a physician, I promise to accept that this oath to my practice of medicine will be continuously cultivated and added to as my experiences are broadened and my mind and eyes opened to novel concepts. This oath is but the roots of my tree of experience and knowledge, whose branches may expand and intertwine endlessly throughout my life in medicine.

I promise,

Jennifer Ann Brooks

As a physician I vow to serve;
to put other people first
I commit myself to a life of learning;
so that I make informed decisions for my patients

As a physician I promise to be honest, moral, and just;
to ensure the patient's trust in me
I will treat every patient as an equal;
so that each feels respected

As a physician I commit myself to the disadvantaged;
to tackle health disparities in the face of minorities
I intend to work in an underserved area;
so that I can be part of change

As a physician I pledge to be culturally competent;
to sensitively and effectively treat my patients
I promise to invest my time and attention when with my patient;
so that a patient-physician relationship can be established

As a physician I will always believe in myself;
to be confident in the shadows of doubt
I will remind myself that I am qualified to be here;
so that I can proceed to succeed when I feel odds are against me

Alexis Perez Rogers

Shelter from the storm
Yearning for a life to live
All are made equal

There are many reasons why I want to become a physician. I want a career that promotes life-long learning, and forces me to continue to improve myself as a student of science. I want to help to further medical knowledge by pushing the boundaries of what we know about the condition of *homo sapiens*. I also want to perform a service to the community, which I believe is worthwhile. I want to be a person who can be relied on in the community and within my family.

Though there are many reasons that contribute to my desire to become a physician, my core belief is that all humans are created equal, and that the greatest/most honorable way a person can spend their life is in the service of others. A physician works to the very core of that belief, pursuing health, while battling suffering and death. It is my simple dedication to that belief that I consider to be my oath as a future physician.

I chose the medium of Haiku to display this oath because, like my conviction, haiku is simple and concise, while conveying a myriad of multifaceted emotions.

Having been given this honor of entering into the medical profession I pledge the following:

I promise to pay due respect to those who have gone before us in this profession and have devoted their time to the teaching of this art. I thank them for their willingness to teach and hope that I will to do the same in the future.

I look forward to working with other doctors and health care professionals to care for patients. I will not be afraid to say, "I do not know" and seek the counsel of others, including attendings and peers.

I vow that I will use all my knowledge and skills to provide the highest quality of care to all patients to the best of my ability. I vow not to let my decisions or recommendations be swayed by personal gain of any kind. And I will remember that part of caring for patients is listening to their stories and addressing them with respect. I look forward to fostering patient-doctor relationships that encourage shared decision-making. I commit to communicating effectively and to always respecting patient privacy.

I promise to maintain my health and live a balanced life so that I can devote my energy and time to serve patients. I hope to use my abilities for the service of the underserved that lack access to healthcare, whether it is locally or globally.

I commit to lifelong learning, not simply in academia, but also in learning to have greater compassion and integrity. In my life, I commit to living in a manner worthy of the medical profession. I look forward to valuing character above all and to not judging based on appearance or first impressions alone.

I value the gift of human life and believe that God has granted us this blessing out of His love. Thus, I promise to strive to heal and care for the sick and injured. And I also promise to help maintain the health of those who are blessed with good health, so these patients may live their lives out fully, pursuing their dreams and goals.

In this race of life, I promise to run it in a way pleasing to God and that includes keeping my pledge faithfully. I know that there will be ups and downs, but, in the end, it will be a meaningful journey of experiencing first-hand the art of healing.

Stephanie Yen

I want to be a doctor. Who gives people hope.
Who makes them happy through genuine smiles, enthusiasm, laughs, understanding, and heart-felt healing.
I value the mind more than the body, and will work to heal both when a patient is ill.
I will put a 100% into everything I do, and reach my personal best every single day. For every single patient.

I value family, and if mine should require my time, such that I cannot put 100% into my work, I will reorganize my tasks, perhaps delegate some responsibilities, so that when I work on something, I am working on it 100%.

I will love my job. I will associate with my patients, and regardless of where I move and where I work, I will see my patients as my community.
It is the individuals that define a group. Healing one individual at a time (both mind and body), will result in a healthier community, will result in a healthier me. That will be my purpose: to better those around me, to better my community and to better myself.

I will not restrict my work to a hospital or clinic. I must work outside of the hospital as well, to maintain a clearer perspective of my world and to help my community in the fullest way I can.
I will take my medical, artistic, journalistic, scientific and leadership skills to parks, libraries, fairs, schools and other gathering areas.
I will organize or strengthen existing organizations that improve the connections within a neighborhood, that inspire students, and develop our youth.

I want to be excited every single day. I will always walk with enthusiasm and a smile. I will be warm and gentle in healing and helping others, aggressive in pursuing my goals, and dedicated to my oath.
Because I want to be the best doctor I can be.

Zainab Saadi

08/25/07

I pledge to dedicate my years of training as a physician to the development of mind and spirit so that I may fully serve my patients' needs. Through academic and professional development, I will work to develop the necessary skills to treat not only the patient, but also the person behind the ailment.

I promise to treat each patient as a unique individual, regardless of race, creed or sexual orientation; to provide each with the best treatment, to the best of my knowledge, serving as a healer, friend and advisor.

I will practice integrity, respect and honor, as I continually strive for excellence in my practice. I will stand as a symbol of life, healing, progress and courage; as a person of faith, imposing not my own beliefs on others but accepting the beliefs of each individual.

I will avoid all that may taint my credibility or practice, focusing not on material wealth but on care for the sick. When temptation draws near, I will remind myself of my commitment to patient care and love for medicine. I will realize that petty dealings are futile and will only leave me feeling empty; that true value arises in the restoration of health.

When days are long and burdens great, I will not compromise my standards of practice. Though challenges will be great, I promise to seek appropriate help when needed. Through my tenure, I will continue to grow both academically and personally. I will learn from colleagues and other professionals, and most importantly, from my patients. I will realize that every patient has a story, and I will listen attentively.

As a physician, I realize my personal and social responsibilities. I acknowledge that in order to serve my community to utmost potential, I first have to prepare myself mentally and emotionally. I realize that my profession entails sadness, but that the fruits of my labor will be great. I know that my desire to effect change and improve human health has prepared me to take on the rigors involved in this profession of medicine.

If I adhere to this oath, may I enjoy countless days of joy, satisfaction and success. May I awake every morning and realize that I am making a difference; that I am working for a greater good and living my dream. May I continually strive for greatness, and realize that the potential for patient care is endless. May I wear my white coat proudly and realize that it stands as a symbol of honor and courage; that I have embarked on a journey of life-long fulfillment and care.

Anonymous

Hippocratic Oath Sonnet

On this lifelong journey, I now embark,

I vow to actively master this art.

To those sick and suffering, I will hark

In hopes of healing unfortunate hearts.

Adherence to high ethical standards

Is a necessary priority

To truly be a respected steward

In this patient-centered community.

I strive to become a better doctor

Keen to impart discerning wisdom; and,

To be health's steadfast, tireless protector

Who always opens a warm, welcome hand.

The opportunities to learn are rife.

I am bound to this oath all through my life.

Edward Chau

I look forward to, in the near future, and have every intention of following the set ideals held in this oath. I vow to treat every patient that I encounter as a friend to whom I wish only the best for. I hope to teach these patients, as it has been taught to me, the importance of caring for one's own mind and body. I hope to learn from my patients anything and everything that they have to teach me. I plan to always reflect on both the good and the bad that my days as a physician have to offer me and to use the experiences to make me a better person and doctor. I will remember why I wanted to become a physician as a young undergraduate student and make every effort to maintain the level of enthusiasm that I emitted in my first week of medical school. I will accept the great amount of confidence that my patients will have in me and I will strive to place that same amount of confidence in myself. I vow to spend my career working to improve the lives of my patients, family, friends and community. I commit to a lifetime of attempting to learn everything, while accepting that no matter how long I live, it is a task outside the realm of human possibility.

Above all, I hope long after I am gone that I will be remembered by the people who knew me as a "good physician," and as a "great person."

Anonymous

I elected to write a short musical sketch entitled "Not in Vain" to serve as a reminder of my commitment to the medical profession. It's a musical interpretation of a passage of the Bible that has greatly influenced my understanding of service and vocation. I hope to play it from time to time to get the melody stuck in my head and the encouragement stuck in my heart. More than encouragement, the words also serve as an exhortation to inspect my motives and quality of work. These verses are one of the most personally meaningful ways I have found to articulate the *why* and *how* behind the characteristics and values upheld by the Hippocratic oath. It deals with questions of death, suffering, and our work to bring redemption in a world that is both deeply broken and incredibly beautiful. My promise is to hold fast to the hope and love upheld in these words and reflected in this music – and Lord willing, will also be reflected in my work these next four years and into my profession.

"Where O death is your victory? Where O death is your sting? The sting of death is sin, and the power of sin is the law. But thanks be to God! He gives us victory through our Lord Jesus Christ.
Therefore, my dear brothers, stand firm. Let nothing move you. Always give yourselves fully to the work of the Lord, because you know that your labor in the Lord is not in vain."

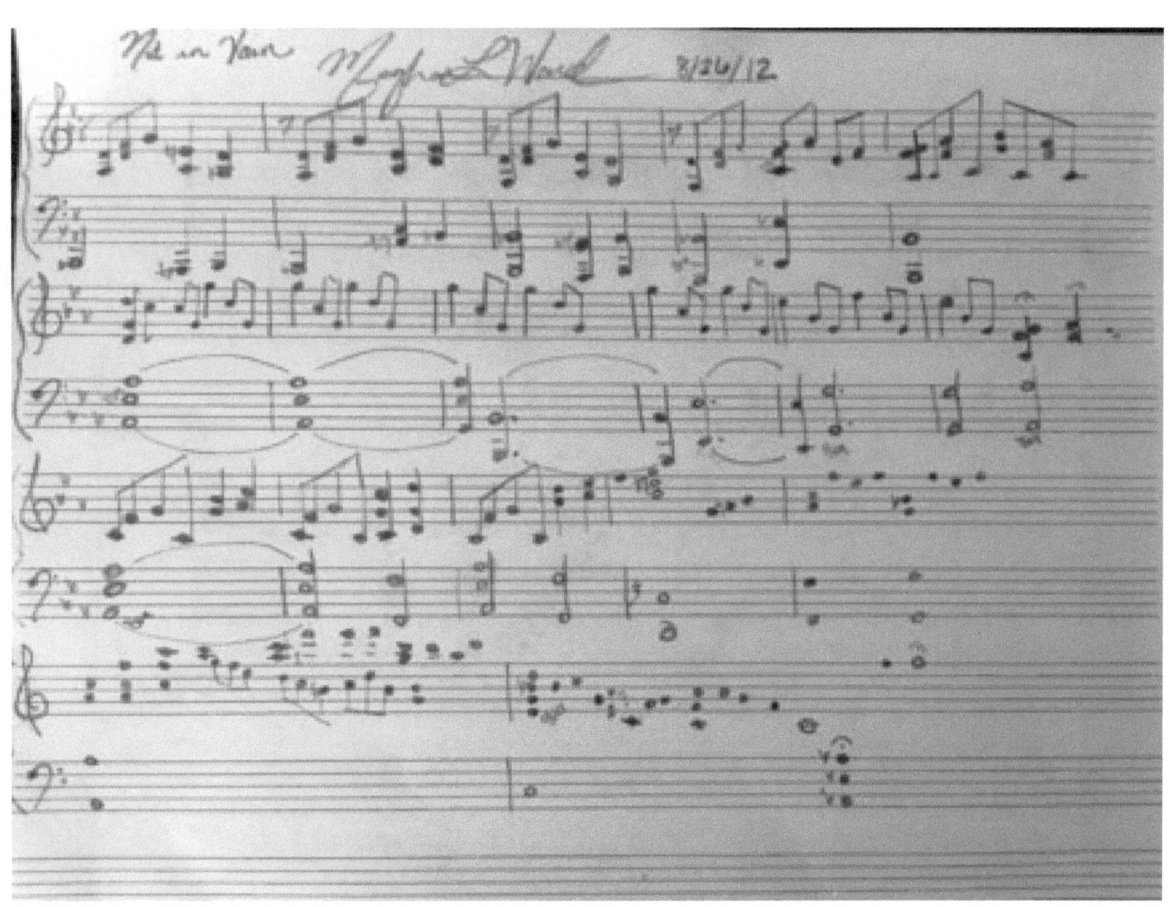

Meghan Ward

Oh Creator, Who placed flesh upon bone

Who knit our inward parts of *Soma* and *Pneuma*

And inscribed in us the image of wisdom and commiseration.

You have known me and named me before my mother bore me.

You have looked upon me and have given me favor with Aesculapius

I will use this favor with utmost diligence and discretion.

When faced with those curious vices of humanity,

I will tell no man no other story but his own.

When Nature's calling runs about like an angry lion, I will show no fear

I will arm myself with rod and reason in defense of Soul and Body.

I will work within the limits of my tarnished image to do no harm.

Though I should fail, may Creation have its undue mercy upon me.

When Nature's beckon comes as a slow and sweet symphony,

I will solemnly submit and allow the completion of its final movement.

And when the final chords resound, I will find my peace in its proceeding silence.

And when my hands grow weary, I will look upon the Architect,

And he will restore my tired soul.

Anonymous

Justin Sleeter

Two Sides of the Same Coin

The look in a patient's eyes when she learns she's dying.
Handing a mother her newborn baby—healthy and screaming.

Telling a family in the waiting area he didn't make it.
Letting her husband know that she's going to pull through.

Blood test shows the cancer has returned right before his 70th birthday.
A scan that shows a nine-year-old is finally in remission.

"There's nothing we can do."
"You're HIV negative."

Removing that eye to prevent the ocular melanoma from spreading.
Diagnosing that mole as noncancerous.

"You should come in to say your goodbyes."
"Hurry in so you'll be there when she wakes up from surgery."

Yelling for the crash cart after his heart stops.
Watching her breathing stabilize.

Mistakenly cutting a nerve that could leave him wheelchair-bound.
Putting in the last screw in his artificial hip.

"Ms. Johnson, please take a seat so we can discuss your options."
"I've got great news for you, Mr. Kaplan."

Handing her a pamphlet on adoption.
Showing her on the ultrasound that she's going to need a second crib.

"I'm sorry."
"Congratulations!"

Medicine. The height of the highs are only matched by the depth of the lows. I swear, no matter where a patient lies on this emotional rollercoaster of a profession, that I will dutifully perform my service to the best of my ability. I will hold all of these experiences, good and bad, with me on the journey that will be my career.

Anonymous

I will not forget... **Passion, Love and Hope**

- I will be passionate about serving others because I believe that my actions, no matter how insignificant they may seem, are part of a larger effort to make this world a better place.

- I am becoming a physician not because it is my goal, but because I believe that the skills that I acquire and the knowledge that I accumulate will be a means to show love to as many people as I can.

- I will treat others without prejudice, regardless of creed, race, education or cultural/social background, and I will look for the best in everyone I meet.

- I promise to use my God-given gifts to help the poor and the socially invisible so that they will feel loved and cared for, and so that they will not lose hope.

- Even though spirituality is important to me, I will not force my worldviews, values and religion onto patients. However, when they are in line with the patient's, I will use whatever I can to give him or her hope.

- I aspire to be a physician who is known for my passion to serve, love for all, and hope for the future of humanity.

- The smiles and laughter of those who I care for have been my source of energy and enthusiasm, and they will be in the future as well.

Mitsuhiko (Miko) Tsukimoto

I swear by my virtue and integrity, to those who have generously given strength,

support, unconditional love, whom I cherish most sacredly,

I WILL BE LOYAL

to not only the profession of Medicine, but also to my values, beliefs, and character

I WILL BE

JUST, GENEROUS, and KIND

to dedicate not only my

mind, skill -------- but rather ---- heart, soul, being

TO PROVIDE holistic and honest care TO HELP

UNCONDITIONALLY

TO ~~ELIMINATE~~

biases and false illusions of

H

I

E

R

A

R

C

H

Y

that society constructs in an effort to blind

TO SEEK

B-O-N-D-S

of

kinship and

=

NOT of

animosity

or

DIS   TINCT   ION

TO UNITE

the TEAM as the FAMILY

thriving,

e   v   o   l   v   i   n   g

under a common desire to universally

DO GOOD

TO LIVE (com)passionately

TO HONOR

tremendous opportunity

&

RETURN > RECEIVE.

I WILL CONTINUALLY SEEK

BETTERMENT

&

INNOVATION

I WILL NOT

be falsely guided by "perfection"

if such a thing may truly exist…

INSTEAD I will strive for

CONTENTMENT

HUMILITY

I will **REFLECT** upon beauty in every interaction | niotcaretni yreve ni ytuaeb nopu **TCELFER** lliw I

TO EXCAVATE

THE SURFACE

---

of "diagnosis" and to holistically render humanized content

TO HONOR the STORY

each individual tells and

ENABLE

the continuance of that narrative

To recognize, but not misuse the gifts and "power" we are given

To lead my life and practice my art in… UPRIGHTNESS and

HONOUR

with…

HONESTY, INTEGRITY, and most importantly, COMPASSION

for the good of the sick;

holding myself far from aloof from wrong,

from corruption,

from the tempting of others to vice

that I will exercise my art solely for the cure of my patients.

WE ARE ALL IN NEED OF HELP.

WE ALL SEEK CONNECTION, COMMUNITY, AND COMARADERIE

to disintegrate boundaries that distinguish us.

To remember

that there is an ART to Medicine

that we must face forward when others turn away,

that we must provide unconditional optimism for those who have no hope

that we must offer our hand, when others may recoil in fear

to gaze directly into the windows of the soul of those who generously offer connection

To open mind and heart affectionately

and both open arms

to both catch and fall into our teammates

TO EMBRACE

not only challenge,

but more importantly the responsibility and honor to do good upon humanity.

Now if I keep this oath and break it not, may I enjoy honor, in my life and art, for all time.

Lauren Maldonado

As a medical professional I intend to practice with compassion. I would like to always put my full effort into seeing my patient's perspective, and to respect that people are (to some extent) products of their environments. Were it not for luck, I could easily be in their position. I accept that my knowledge is limited and because of this, every aspect of my patient's life experience demands respect. I will never stand in judgment of my patients.

I will always put my patient's interests first, above monetary or social gain. This is because I believe that someone needs to exist as an advocate for my patients. If I do not then who else will? I will treat my patients with respect and I will not tolerate disrespectful (much less, dangerous) behavior towards a patient from a colleague or coworker.

I will do my best to stay up to date on the pertinent technologies and research developed in my chosen specialty so that I can provide my patients with good counsel. I will not speak with false authority on a subject that I am uninformed about, or risk my patient's safety for the sake of ego or social gain. I will make every effort to explain to my patients what interventions I plan to take and why. Knowledge is power and every person deserves knowledge and therefore power over their own physical body.

I will stay up to date on my skills in order to provide my patients with the best care I am capable of providing. I will acknowledge my patients as people, deserving of emotional support and comfort, and I will make every effort to support their mental state as well as their physical wellbeing. I will do my best to provide my patients with an environment that alleviates their anxiety. I intend to practice with a demeanor that provides my patients with a sense of confidence in my skills as a physician and that promotes trust and a sense of confidentiality. I will respect my patient's autonomy and right to privacy, unless I determine through rigorous and logical analysis that infringing on either will improve my patient's quality of life. Then and only then will I ever infringe on those rights and with full knowledge of my actions.

Anonymous

i am no longer who i was before and thus i the things that i think do and act and say and feel will be set a different level a higher bar a new step a higher step a different perspective vantage viewpoint on life and duty and responsibility

oaths are not written said made for no purpose it is something i vow and with great power comes great responsibility and i will not make light of it even though the oath is only words uttered i feel it as a cataclysmic shift a momentous moment the precipice the turning point the edge the end of the old and the beginning of the new

i hope to never forget how i feel at this moment my thoughts my feelings my being my essence the happiness and elation and euphoria and joy and pride to be called among the few because this is a truly a calling not a job not a profession not a nine to five not even a life's work but a calling and inner impetus to work for others and to give of myself

above all i will remember how i felt as a patient sitting in a cold sterile room scared and frightened and alone and with fear pressing in on me on all sides shame and fear and sadness and a complete emptiness of chemicals in my brain and wanting to cry out for help for a helping hand for anyone to be in that room with me someone to share my pain but there was no one and there was only a man doctor who did not offer me a comforting word who saw girls like me every day many a day hundreds a month and exponentially more in a career in a lifetime and couldn't feel enough to reach out and make me feel grounded to this world or ask me a simple silly small talk question that most of us scoff at but that would have gladly made me do something anything to distract even for a second from the pain

as a family member watching another member of my own being of my own flesh and blood lying on a cot helpless and barely able to move and me feeling so helpless and scared and not knowing and not understanding

but i am not bitter but rather my oath is to remember these moments so clearly in my mind so that i will impart what i knew and what i experienced as a patient as a daughter a granddaughter as a human as a simple living being existing in this world and in this life and no matter how tired how sad how angry how emotions how anything that comes to rest on my shoulders i will bear them and i will comfort those in need because they have put their trust in me put faith put life put parts of themselves put their desperation and fear and pain and my oath THE oath that i give is to see from their eyes to feel from their being make their pain my own own their burdens as my own burdens

If I keep this oath faithfully, may I enjoy my life and practice my art, respected by all humanity and in all times i have been given a gift divine and human and real and incarnate and precious and the knowledge i am imparted i am gifted with i will prize everything cherish every second and may i spend my entire career calling profession life time on this earth and time beyond that time and to infinitum in debt to the higher powers that be may i spend that time living to my oath with every fiber of my being and giving with everything of myself to the service of my patient and to keeping our oath the oath my oath

christina gainey

My Oath

I do not know what challenges lay ahead of me, but I will face them head-on.

I will care for those I know and those who are the most foreign to me with the same degree of compassion and respect.

I will work for the betterment of all mankind, but I will be satisfied and grateful for the privilege of caring for and uplifting just one.

I pledge to take care of myself enough to take care of others.

When I am consumed by doubt, when I am overwhelmed, when I want to give up, I pledge to push through, for your sake and mine.

I will learn to do everything in my power to help my patients without being crushed when there is nothing I can do. I will recognize the limits of medicine and the limits of my own humanity.

I aspire to not let social class, race, or any other man-made construction create a divide between my peers and my patients and myself.

I will look into your eyes and know that we are the same.

Elizabeth Schanler

I have always done my best to live my life by the simple code of honor, fairness, kindness and truth. While simple, those four words carry significant meaning to me, lay the foundation of who I am and guide my actions. By choosing to become a physician, I have chosen to spend the majority of my time serving society; using my mind and body to make others' lives better. To allow me to perform my duties, society will give me extraordinary privileges. To this end, those four words become infinitely more important.

As I travel down the path towards becoming a physician and practicing the art of medicine, I promise to always remember that I am serving my patients, they are not serving me; to be humble; to act with honor; to treat others fairly and justly; to not discriminate, in giving treatment or otherwise, against anyone regardless of the circumstances that I find them in; to hold myself from corruption and from the allures of fame and fortune when they would interfere with the proper and thorough treatment of my patients; to hold secret any information revealed to me in confidence; to respect my patients wishes; to ask for help when needed and to give help when asked. In short, I promise to deserve the trust the trust that society is placing in me.

Being a good physician starts with being a good person and I promise to work hard on being both.

Anonymous

I swear to all that I hold true in my heart, that I will strive to fulfill the duties set forth in the following oath.

I vow never to allow a patient's religion, race, sexual tendencies, and moral or social standing impact the quality of care that I will strive to deliver.

I promise to always train others in the practice of medicine, independent of personal gain or status.

I will at all times diligently strive to ascertain the most current facts and techniques of treatment, such that I will never deprive my patients of the most modern remedies to aid in their suffering.

I vow to always respect the doctor – patient relationship and remain free of any intentional injustices with regard to mischief and sexual relations.

I commit to always maintain the patient's best interest in mind when prescribing a treatment program; I will not let personal gain, ambition, or moral beliefs, alter a treatment protocol.

I will never knowingly administer or remove a treatment which has as its sole purpose the role of reducing an otherwise healthy person's livelihood.

By fulfilling my oath above, may I continue to practice the art of medicine; if I deviate from my sworn oath, may all my privileges as a physician be revoked.

Aaron Bond

My Pyramid to Medicine
Adapted from John Wooden's Pyramid of Success

Greatness
I vow to remember that a good physician treats the disease while a great physician treats the patient

Poise
I vow to be myself and not be thrown off by good or bad events

Confidence
I vow to believe in myself and know that my work has prepared me for anything I face

Morality
I vow to treat every patient equally no matter his or her race, religion, socioeconomic status, or sexual orientation

Learning
I vow to continue to develop and expand my abilities every year and be a lifelong learner in the art and science of medicine

Community
I vow to give back to the medical community through research, pro bono service, and mentorship of future physicians

Professionalism
I vow to be a punctual, respectful, and accountable professional every day

Alertness
I vow to be in the present moment and to respect the time of my patients and the hospital staff

Courage
I vow to have the courage to step up if I see, hear or witness something that I know is wrong

Confidentiality
I vow to maintain a patient's confidentiality and respect the doctor-patient relationship

Industriousness
I vow to diagnose and treat each patient to the best of my abilities for every single patient that I see

Empathy
I vow to realize that every patient has a story and support the person behind the patient to the best of my abilities

Altruism
I vow to place the needs of my patients before my own and be an advocate for their care

Cooperation
I vow to work together with other physicians and hospital staff to provide the best care possible for our patients

Enthusiasm
I vow to be passionate and excited about medicine and realize how blessed I am to be part of such a profession

Alana Munger

It is tradition as a clinician
To make a Hippocratic oath rendition
Thus with ambition I state my definition
Of what it means to be an ethical physician

I aim to treat, never to cheat
My practice shall always be void of deceit
I must be discreet with everyone I meet
And truly earn my place among the elite

In order to protect, detect, and disinfect
Patients can expect every option to be checked
Yet there are times that I will be incorrect
Medicine is not something one can perfect

Throughout my career I'll need the help of a peer
I may not know an answer, I may be unclear
I will never fear to ask help from those near
For the sake of our patients we must cohere

I must know the latest in medical news
Whose new drugs are good, what methods to use
And thus I'll peruse many different views
So without excuse I'll know the best option to choose

So shall I err against this that I swear
May I forever live a life of despair
For no matter who, what, when, why, or where
As a doctor it's only fair I provide ethical care

Merrick Lester Bautista

"Learning does not make one learned: there are those who have knowledge and those who have understanding. The first requires memory, the second philosophy."

I value the importance of knowledge and understanding the science of medicine. However, I vow to not only obtain the appropriate amount of knowledge, but to also gain a sense of understanding of how to practice medicine. Throughout my medical education, I hope to learn how to practice the science of medicine from my professors and the art of medicine from my patients.

"God wants Man, whom he has created and in whose heart he has so profoundly entrenched a love for life, to do all he can to preserve an existence that is sometimes so painful, but always so dear to him."

I can understand the amazing phenomena of every person's desire to live and avoid death. As a person, I too will be driven by my passion to live no matter how hard the circumstances. However, as a physician, I must also do all that I can to preserve the lives of others. I understand that this may be very hard and painful, but it is my responsibility and it will always be of utmost importance.

"To save a man's life, to spare a father's torment and to protect a mother's feelings is not a good deed, it is an act of mere humanity."

I vow not to view my role as a physician as job or employment. It is my passion and calling to help anyone I can to my greatest ability. It is not my responsibility but rather an embodiment of my love towards others.

"Only someone who has suffered the deepest misfortune is capable of experiencing the heights of felicity...Until the day when God deigns to unveil the future to mankind, all human wisdom is contained in these two words: 'wait' and 'hope'!"

For some reason or another, this quote saved me from life's many struggles. I will continue to use these reassuring words to my own hardships. I promise to not leave my patient's side until I have given them hope. I understand that healing takes a lot of time and patience. During the healing process, keeping hope alive is of the utmost importance, especially in times of uncertainty. I believe that at least uncertainty means the continuation of hope and I vow to never forget the importance of hope for my patients and myself.

Quotes from: *The Count of Monte Cristo* by Alexandre Dumas (my favorite book)

With deep sincerity,

Neda Roosta

Shattering the Glass Floor

The painting is about shattering the masks we all wear including the mask of privilege in order to shatter the barrier of comfort so that we might help heal and learn from those on the margins. On the bottom left is the Hippocratic Oath in the original Ionic Greek and on the top right is Matthew 25:31-46 in Hebrew. The fire represents initial sacrifice and suffering, but later thriving and awe by both the patient and physician.

Warren Yamashita

I give my unwavering commitment to the following covenant:

I acknowledge that in gaining entry into the practice of medicine, I am bestowed with a special set of responsibilities. I recognize that by the very nature of my profession, my relationship to patients is a power relationship. In seeking my services, my patients are in an especially vulnerable position, significantly dependent on my expertise and judgment. As such, I vow to direct my professional efforts in service to my patients, always holding their well-being above all other considerations.

I will not discriminate under considerations of race, gender, religion, culture, language, class, political views or any other forms of tribalism that arbitrarily divide the human race. I will treat every patient equally, while simultaneously accounting for each patient's particular individual and social considerations. My interventions will be unwaveringly guided by logic, reason and the most current scientific findings, never neglecting all relevant ethical considerations. I will make every effort to prevent illness, as proactive prevention is always preferable to reactive treatment.

In recognizing my role as a patient advocate, I will not be restrained to operating within the biomedical model of medicine. In addition to the chemical, biological, physiological and personal considerations that guide my interventions, I will always recognize and consider the socioeconomic, geopolitical, ecological and environmental determinants of health. I will not be constrained to treating patients at the personal and molecular level, but will extend my advocacy efforts to the population level when warranted.

I will not be tempted into engaging in personal pursuits of social status or income. Not only will I refrain from engaging in political activities for the purpose of profiteering at the expense of the public, but, as a patient advocate, I will actively confront and attempt to thwart these endeavors.

I will respect the justified forms of hierarchy under which I operate, be they legal, institutional or professional. However, the ethical basis of my behavior will not be derived from any law, institutional policy, or code of ethics - any of which are subject to the biases of the particular historical context in which they were created and can often become anachronistic. Rather, the fundamental ethical guiding light of my behavior will always be derived from the profound lucid realization that the entire purpose of my professional existence is service to my patients. My commitment as a physician is not to the organized profession of medicine, not to the institution under which I perform my work, and not to the government who regulates my activities. My professional commitment is, and will always be, to the patients and communities I serve.

I acknowledge and accept that, ultimately, the essence of what it means to be a physician is to restore and maintain patient autonomy; autonomy to make reasonably well-educated decisions about their health, and, when possible, autonomy to carry on in their pursuit of happiness, partially if not completely free from the shackles of illness and disability. In closing, as a physician, my aspirations are derived from a set of twin pursuits - one intellectual and one moral: 1) I hope to better understand how the world works, and 2) I hope to use this knowledge to help lessen the suffering of others. If I can accomplish these goals, I will have met my professional obligations and fulfilled my purpose as a physician.

Ricardo Padilla

A SOLEMN OATH

A solemn oath I have taken,
I vow to keep it
To embrace where I am in life, where I am going
Reflecting

A white coat I put on
I vow to wear it
To understand what it means to be a doctor
Pondering

A unique life I have chosen
I vow to honor it
To overcome the darkest days
Persevering

A challenging road I have ahead of me
I vow to walk it
To take the unbeaten path
Exploring

A mind full of uncertainty about parts of my life
I vow not to ignore it
To be cognizant of my heart
Listening

A true excitement meeting my new colleagues
I vow to appreciate it
To feel happy, relaxed, relieved
Celebrating

A realization that I found what I love
I vow to grasp tightly onto it
To not overthink the reasons I am here
Smiling

A vision of the future
I vow to realize it
To be myself
Doctoring

Anonymous

My Oath...

My mind, plagued by questions, novel and queer
Ponders my worthiness for this mantle.
In mine hands, a patient's faith, so sincere,
I must grapple fear to disentangle.

I don a coat of luminous virtue,
proclaiming strength, wisdom to other's eyes.
But my heart encased in worry - a statue -
Quakes, shatters, thinking of faith's demise.

Incongruity of uncertainty
With an education steeped in detail,
in candor must be harnessed ardently.
Its presence reveals a wish to avail.

Faith, duly placed, while seeking remedy,
I hold warily, lest lose my empathy.

Kirt Gill

I wrote a sonnet (or rather tried to write one) expressing what I believe to be the most important factor in a long career in medicine: Remaining empathic and cognizant of the intimacy present in a physician-patient relationship.

I vow to approach my medical education with enthusiasm, honor, and determination. I promise to dedicate myself to my studies and to work to the best of my ability every day that I can. I will strive to remember that each patient is a person with a story, not just an aggregate of symptoms, and that working with patients is both a privilege and an essential element of my medical education. I commit myself to treating each patient that I meet with respect, compassion, and sensitivity.

I recognize that medicine is based on teamwork, not competition. I will remember to ask for help from my instructors and my peers when I need it, and to help others when they are in need of my assistance. I vow to be honest with my patients, my colleagues, and myself. I recognize that becoming a doctor means making a commitment to lifelong learning, and I vow to stay informed about medical developments and discoveries throughout my career. I promise to avoid becoming complacent in my medical knowledge and to recognize that there is always something new to learn from my patients, co-workers, and colleagues. I recognize the fears and anxieties that I will have throughout my medical education and career, and I vow that although they may trouble me, I will not allow them to rule me.

I realize that being a physician is a great honor, and I promise to work to earn the respect and trust of the society that has given me this honor. I vow to govern myself as a professional, to behave, speak and dress with decorum and honor. I promise to work as hard as I can to be knowledgeable, professional, honest, and compassionate. I vow to remember that from my first day of medical school, I will serve as an ambassador of the medical profession to everyone that I meet. Therefore, I will work my hardest to live up to and better the reputation of the medical profession. I hope that through hard work, I will find my career to be exciting, fulfilling, and rewarding.

Anonymous

It is my duty to act with honor constantly. I vow to those that came before me that I will act with the highest values in mind, and that my character will remain formidable especially during the times of utmost temptation.

"The measure of a man's real character is what he would do if he knew he never would be found out." ~Thomas Babington Macaulay

I commit to acting only with my patients' best interests in mind, and do nothing to jeopardize their health or well-being.

"Love all, trust a few, do wrong to none." ~William Shakespeare

I will always value how I care for an individual over fame, fortune, or any type of benefit.

"Try not to become a man of success but rather try to become a man of value." ~Albert Einstein

I vow to act with the utmost integrity when interacting with my patients, my colleagues, and generally everyone I interact with daily.

"If you have integrity, nothing else matters. If you don't have integrity, nothing else matters." ~Alan Simpson

As physicians we will inherit a great deal of power. I will protect my patients' secrets as if they were my own. I vow to never abuse the power that is inherent to all physicians.

"Nearly all men can stand adversity, but if you want to test a man's character, give him power." ~Abraham Lincoln

I vow to help anyone in need. I vow to serve the underserved throughout my career both domestically and abroad. I vow to contribute to the improvement of global medical care.

"Be the change you want to see in the world." ~Mahatma Gandhi

"The only thing necessary for the triumph of evil is for good men to do nothing." ~Edmund Burke

I used some of my favorite quotes to address the parts of the Hippocratic Oath that meant the most to me. In life, I always try to do the honorable thing. I act with integrity and having a strong character is extremely important to me. These values will be no different when I apply them to medicine. Additionally, I have a passion for service and one of my favorite quotes (Edmund Burke) has been particularly meaningful to me over the last three years. I intend to pursue this type of service during my medical career.

Michael Toboni

I forgo the primrose path in dedication to service in medicine, to want not but for the good of my patient, to seek not but what will bring healing to those in need of its warmth. I swear upon my integrity as a member of the human race that I will embrace the full duty and obligations of my title in such that it will be my life's work, forever in pursuit of excellence; that I will strive to seek what is right though the path may be fraught with resistance; that I will conduct myself in a manner worthy of my title, and if I should fail to do so, I will without reservation relinquish the title of which I am undeserving; that I will act with compassion and empathy, recognizing that every life is worthy of appreciation; that I will work with my whole heart, honestly, faithfully, dutifully, and diligently; and if I should not break my vow, may I find eternal joy in sharing in the lives of those for whom I shall care.

It is through the grace of God—His sovereignty and purpose—that I owe my current identity and also my calling, as it pertains to my future as a physician. Thus, before Him do I promise these things:

I dedicate my life to follow in the footsteps of the Great Physician and to regard all others in the medical profession with the same reverence and humility. Those who teach me, I will treat with utmost respect; those who are my peers, I will esteem as equal to or greater than myself; those with whom I work, I will value as priceless.

To the best of my ability, and further still by the strength of Him who has called me, I will serve those who are placed under my care. I will treat every person as one who bears the image of God, and thus as an object of infinite worth. In all situations, I will remember that I am treating souls and not simply bodies. By these principles, I will work for the best interests of those in my care, striving to hold myself blameless and pure from wrongdoing of any kind. I will practice my art for the purpose of healing alone and will hold in strict confidence everything with which I am entrusted.

At all costs, I commit to serving the least of these. I will attune my ear to the cries of the widow, the orphan, and all others who are helpless. At the focus of my art, I will place those who are poor, lonely, oppressed, and suffering, whether here in California or across the world. I will not ignore them, take advantage of their situation, or mock them.

In every area of my life and art, I commit to acting with integrity that is beyond reproach and consistent with my identity as a child of God. I will endeavor to show compassion, grace, and mercy to all those I encounter, whether peer or patient; to demonstrate love that reflects the far greater love, which has been bestowed upon me.

Finally, I will always remember that I am treating the whole person; that my most valuable service is not physical healing but a gift of hope that is beyond this world; that it is the Spirit who does the heavy lifting. By the grace of God, I will stay faithful to my commitments, and may His healing power be made known to all those whom I serve.

Michael Tan

As a Future Pediatrician

I promise, not only to be loyal and true in the practice in medicine,
But, as a future Pediatrician, to think always about my patients,
For they are a vulnerable population, and cannot always make their voices heard.

I promise to use my powers of observation, my patience, and my skills
To diagnose them, with respect and consideration,
Remembering that even at such a young age, all children are people, too,
With real thoughts, fears, emotions, pains, and joys.

I promise to think of their parents, some sick themselves with worry,
Some despairing over chronic illnesses,
Some struggling to make ends meet and do the best for their children with what they have.

I promise to bring comfort, and education, so parents can make their best, informed decisions.

I promise to be honest and true and to look for honesty, in turn, from the parents as they relate their children's histories.

When things don't add up, I will go that extra step to see what must be done.
I know not where life will take me, but I hope to be thoughtful, capable, and kind.
Though I may fall short on some days, I will never stop trying.

This is my promise.

Laura Anne Lewis

I vow to remember the passion I feel on this day.

I realize that I will face many challenges but I promise to meet them with my whole heart and with the best of my ability.

I commit to a life of serving others, of remembering that each man, woman, and child I will treat has a family who loves them and thus to treat them as individuals, not as a list of symptoms.

I let go of my fear, of any feelings of inadequacy, and I trust that I can do this.

I remember that I am not in this alone, and that I can count on the support of my peers.

I promise to ask for help when I need it and to take care of myself so that I can take care of others.

I intend to keep balance in my life, to remember my loved ones as well as my patients, and to remember that, without my family and friends, and, first and foremost, without God, I could not and would not be the person I am today.

I trust that they will see me through.

I vow to try to see good in each person, to find compassion for them no matter what their situation is.

I know that there will be days where I am exhausted, moments where I feel that I simply cannot go on, that I don't have it in me.

And when those times come, I vow to keep going, to remember why I have chosen this life for myself, to remember that if I can make a difference in the life of just one person, it will all have been worth it.

I trust.

I commit.

I believe.

Melissa Hikari Luttio

I promise to commit, entirely, my mind and body to the practice of medicine whenever I wear my white coat. I vow that I will make every possible effort to set aside my own personal problems, prejudices, and shortcomings so that I may conduct myself as the consummate professional for the benefit of my patients. I shall treat each individual who comes under my care as more than a collection of symptoms and categories, but as individual men, women, and children deserving of my respect, my concern, and my most sincere efforts.

While practicing medicine, I will dedicate myself exclusively to the health – physical and emotional – of my patients. I vow to deal plainly with my patients, providing each person with complete and honest information so that they do not fear what they do not understand. I vow to nurture hope wherever I find it, however small, so that it may serve as a light to my patients in a world that suddenly seems utterly dark. I vow to remember how it feels to be sick, so that I might maintain the tenderness and compassion that the sick so desperately crave.

I will practice my craft with the utmost respect for those who came before me, and those who will follow. As such, I will continue to study the art and science of medicine long after I leave school and continue the honorable traditions of my predecessors so that those who come after will enjoy the same esteemed position that I inherited.

I vow to remain faithful to the principles of this oath for so long as I call myself a physician.

Anonymous

Hippocratic Trope

I have written this song, which is intended to be set to the tune of
Coolio's *Gangsta's Paradise*.

As I walk down San Pablo, in the Shadow of Keck,
I take a look at this Oath, and realize its weight and depth,
'Cause I've been practicing MCATs for so long,
That it's easy for me to forget that it's just begun
But I ain't never saved a life, or even helped to treat it
Me be keepin' people breathin', boy I can't believe it
You better watch how you're readin' your patient's symptoms
Or you and your teammates may mistreat the illness
I really hate to rant, but you gotta know
We're held to higher standards now, better heed my flow, youuu've
Signed a social contract to keep health and wellness intact
Fact-based treatments for your patients, life-long learning expectations

I swear by all the gods, to hold sacred all my patients' lives
I'll respect all their rights, keeping secrets lest they threaten lives
A Lifetime of learning's nigh, so you'd better start Netter's tonight
I swear by all the gods, to hold sacred all my patients' lives

Look at the situation your patient's facin'
Help them lead a normal life; they should know it's their right
So you gotta be down with your whole team
Treating patients well means working well with colleagues
You'll be educated to save lives on the line
Life or death is not a game, negligence is a crime
You're a stressed-out student in debt, sippin' coffee
And your brains are so fried, your notes are getting' sloppy, youuuu
Slept through genetics, didn't prep for statistics,
Couldn't listen, blurry vision's got you missin' cell division
You're a first year now, but will you make it to MS2
Your big sib says it's up to you

Tell me why are we, expected to be
Here at 8 a.m. every morning

I swear by all the gods, to hold sacred all my patients' lives
I'll respect all their rights, keeping secrets lest they threaten lives
A Lifetime of learning's nigh, so you'd better start Netter's tonight
I swear by all the gods, to hold sacred all my patients' lives

Lucas Struycken

I solemnly swear by Apollo the Physician, the God to whom I have prayed and whom I have thanked for my entire life, and all the Gods and Goddesses of the religions of the world, that I will wholeheartedly commit myself to the profession of Medicine.

Becoming a physician is not just a professional role; rather, it will embody every essence of my life and my character. As such, I will extend kindness, integrity, and compassion to my work and interactions with colleagues and patients alike, just as I do in my personal life with beloved family and friends. I promise to hold myself to the same high ethical standards, meaning that in whatever context in which I practice – hospital, community, or home – I will remain free of corruption, bribes, and all other potential vices that I may encounter along my path. I vow to prioritize the physical and emotional well-being of my patients above all other factors, including and especially financial compensation. I will strive earnestly to listen to my patients in a way that is respectful and free of all judgment. When my patients share personal information about their lives with me, I promise to keep it confidential.

As a physician, I commit to use the knowledge and skills I gain to contribute to the well-being of my family members and community. I follow in the esteemed footsteps of my late father and paternal grandfather, physicians whose deep caring and empathy for their patients serve as my fundamental inspiration. Remaining aware of massive healthcare disparities that exist in all communities across the world, I will dedicate time to sharing my expertise to improve the health and impact the lives of those who are neglected for reasons related to income, age, cultural differences or anything else. Meanwhile, I promise to give back to the profession of medicine through mentorship of the next generation of physicians.

Most importantly, I solemnly swear to do no harm. Should I adhere absolutely and completely to all the tenets of this oath, may I have a long life and medical practice full of true honor, respect, and fulfillment.

Susan Joan Mauriello

I vow by all I hold dear to keep this oath:

To first do no harm, to take under consideration the full breadth of illness; not only the physical but the mental and emotional aspects as well.

To have the patience to listen to those who I'm trying to treat as well as my colleagues; to not rush to treat, to take the time to formulate a plan that allows all to work together as a team.

To imagine what it's like to be in the shoes of all who are affected by illness; the patient, their families and loved ones.

To treat all that come my way without judgment of their choices and show them the utmost respect.

To follow rules, guidelines and laws, to never be led by temptation to stray into criminal territory, even when there are no eyes upon me.

To protect my patients from harm, whether it be from themselves or those that seek to abuse them. To let them find refuge in our relationship so they can share with me all their burdens and trust that I won't tell another.

To work every day of my life to become a better physician; to learn something new, perfect my craft, and constantly improve all my skills in art of healing.

If I can work at all of these things in return I'll receive a lifetime in medicine that is fulfilling, rewarding and challenging. A lifetime in which I will have given willingly of my time to make a difference. A lifetime that will leave me infinitely grateful for all I have helped and all I have learned from them in return.

Anonymous

As I embark,
On this new journey,
I vow to uphold
These values that I hold dearly:

For every patient
For whom I care
I will give my best,
That I swear.

With honesty and kindness,
I will speak.
And give relentless care
To all who seek.

To those who feel
Weak and alone,
I will stand by them,
Make them feel that I am their own.

I will treat
Both body and soul.
And care for the patient
And the family as a whole.

The latest research,
I will follow and read.
Collaborating with my team,
We will improve and lead.

Even when
Challenges do face me,
I promise to act,
Always with integrity.

Above all,
I will do no harm.
Nor ever act in a way
To cause others alarm.

I understand
This may not be easy.
But no nobler endeavor
Exists to me.

Let it be known
That from today,
To better others and better myself,
I promise to always act in this way.

Rupan Bose

The Words

I swear to those I serve, and to my true self, this oath to the practice of medicine. And, in doing so, I pledge my life to the honing of my medical skills and the passion with which I carry out my profession. Wherever I go I shall carry this obligation with me, therefore all my days will be spent on a mission to aid others. I will resist sloth and cynicism that would tear me down if I were not vigilant. I will not lose the love I have for my patients and the healing arts. I will not use my knowledge to manipulate, but to educate others so that they may be empowered in matters of their own health.

Anyone can harm another person, but few can mend - for that is the greater challenge. I will forever choose the latter. In helping others, I will never lose sight of my own well-being; I will be an exemplar in addition to being a practitioner. Those who confide in me as patients shall find my confidence unyielding. And in my care, they shall also see that no one has loved his craft more than I do. If I take these words as more than just words and write them plain on my soul then I shall truly be a healer of man. I shall answer my highest calling.

Anonymous

To my patients, I vow to	To the profession of medicine, I vow to	To myself, I vow to
practice the art of healing with a commitment to cura personalis;	believe in and trust medicine;	carry out my life as I do my art;
listen and believe you;	practice in the purity of white;	commit myself to all factors of health;
teach health and wellbeing;	motivate and collaborate with colleagues;	work beside, and not above, other health care professions;
protect your lives and information;	advance the delivery of medical care;	choose honorable actions;
remember first your names and personal stories and second your diagnoses and outcomes;	learn and improve our science;	acquire integrity;
	defend our profession;	value life and love;
recognize the dignity in each;	represent our patients' voices;	seek happiness;
do my best, always; And be a woman for others	do my best, always; And be a woman for others	do my best, always; And be a woman for others

Anonymous

I appeal to the sense of integrity and dedication instilled in my by my parents and community that the following will be true:

As I enter into this privileged profession, I will remember that it is a lifestyle that must influence and positively affect all aspects of my life. I will continuously remember that, in order to be of the best service, I must be the best that I can be. I vow to approach medical knowledge and skills with the reverence, focus and diligence that they deserve. I promise to let the potential of my mind, my physical capability and my compassion come to full fruition. I will never end the day knowing that I could have tried harder.

I promise to respect the trust that my patients extend to me. I vow to first view them as independent equals with the profound capacity to make rational decisions about their health, regardless of whether I agree with those decisions. I will not view my patients as an illness, an insurance card, or as a slot on my schedule. I will care for them with the respect, competency, and empathy with which I would treat my own family. I will listen, ask questions, offer advice, and give resources; I vow to have a dialogue instead of a monologue. I promise to respect my patients in life and in death.

My skills will be of service to my entire community (locally and globally). I will strive to provide education and spread awareness to combat issues that I see. I will be a proactive member of my community. I intend to understand the culture, society, and traditions of my community and use my creativity to come up with culturally competent health solutions. Finally, I commit to supporting the other health professionals in the medical community. I vow to remember that every professional in a healthcare setting is an important member of the team and that there is no shame in asking for help. I promise to use my knowledge and skills to the best of my ability to leave my patients, my community and the medical profession better than I found it.

Pooja Jaeel

TWO-HANDED MEDICINE

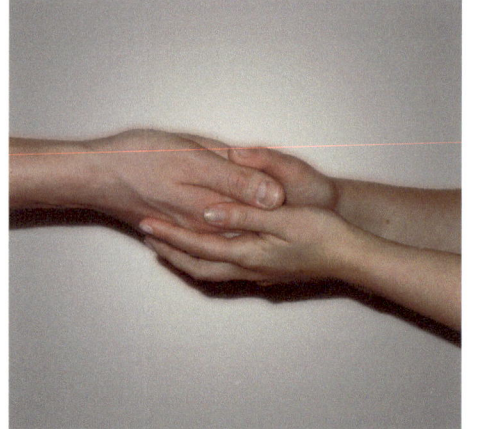

To the Keck School of Medicine and the local community it serves; to my classmates, family members, and friends; to the United States of America and all of its people; and to the world at large with all of its diverse cultures and nations, I pledge to practice two-handed medicine and all that it entails. What, then, is two-handed medicine? Perhaps I should elaborate.

The first step in understanding this concept is to look at the image above. How do you feel when you look at it? I feel comfort, warmth, and security — a sense of calm and reassurance. It shows two people *holding* hands (two-handed) rather than simply *shaking* hands (one-handed). This distinction is important. Hand shaking is a means for extending greetings, initiating relationships, and making agreements. Hand holding, at least for me, carries a different set of meanings. It is for quelling fears and calming nerves; for providing comfort and offering reassurance; for sharing emotions and fostering understanding; and, ultimately, for letting someone know that you care both for and about him/her.

In medicine, then, we could say that hand shaking is a prerequisite, a baseline for individuals providing healthcare for patients, a common courtesy. Hand holding, however, is what defines medicine as a profession and makes a physician a physician. I mean, let us think about this idea. Hand holding is not automatic. It requires action, thought, intention, and, above all, compassion; it requires willingness to be kind and patient, to stand with patients in their most vulnerable moments, and to listen to their stories. And such hand holding serves to build and strengthen a relationship of mutual trust and respect.

This relationship of mutual trust and respect has as its backbone a social contract which includes certain expectations to which society holds the medical profession, among which are acting with integrity and good will, upholding ethical principles in research and clinical practice, putting the patient's interests above one's own, and protecting patient information and confidentiality. This social contract compels physicians not only to execute these actions personally but also to hold their peers (i.e., other physicians) to the same standards to ensure that the profession remains true to its mission of just service. Furthermore, in order to carry out this mission of serving society, physicians must advocate for the well-being of patients by working to address the multifarious factors which impact health and wellness.

Given what I have described, I can now make my pledge. I pledge myself to society and the field of medicine, including all of the elements that I have described in this oath, and I vow to always practice two-handed medicine as I serve the public as a future physician.

Justin Trop

I swear by Apollo the physician, and all other entities holy and healing, that I will dedicate myself to both the profession of medicine and its persons by which it affects. No longer shall I work solely for my own interest, but I vow to work first and foremost for the benefit of my patients.

I will strive to maintain the delicate balance of science and art within medicine. Each patient whom I encounter, I will value as an individual, with his or her own life story and unique personality and spirituality. And with each patient case, I shall take hold of as an invaluable opportunity to not only gain scientific knowledge and expertise, but also continue to perfect the art of the doctor-patient relationship.

I will treat my patients with proper medications and procedures, but also explain to them how they must take care of themselves outside the visit. I promise to promote the significance of preventative medicine – be it lifestyle, nutrition, proper use of supplemental medications, and any other avenues of improving one's quality of life physiologically, emotionally, mentally, and spiritually.

I will not take advantage of my patients nor the medical system itself for personal gain – be it compromising private health information, writing improper prescriptions, or other harmful and unjust actions against the patient, the profession, or the health care system.

I promise to maintain honesty and integrity throughout the medical community. I will stand by my colleagues and support them, as I would expect them to do the same for me. May I encourage them frequently and rebuke them when necessary.

As I commit to this code and integrate it into my life, may I be blessed with prosperity and success. Shall I ever break this promise with intention, may I face the consequences that guide me back towards the path by which this oath proclaims.

Jimmy Mao

I swear that throughout my studies and in my later medical career, I shall adhere to the following standards.

- I shall uphold the moral and individual human rights of my patients, colleagues, and all those affected by my actions to the highest regard.
- I shall strive to use the best of my abilities and talents to make the most positive impact on society.
- I shall remember that my role as a clinician and as a researcher is to provide the best possible service for my immediate patient and for those affected by my work.
- I shall honor, respect, and appreciate my instructors, peers, and medical colleagues in the team environment of medicine.
- I shall view my career in this field as an amazing privilege and be continually grateful and humble for the opportunity to do good.

If I follow these standards, may my practice of medicine be successful and may the world be a happier and healthier place because of my efforts.

David Mittelstein

I intend to treat my patients with knowledge and diligence and make the best possible effort to heal them. I will be respectful of their concerns, cultures, lifestyles, and privacy so that I may better understand how to treat them and not simply relieve their temporary symptoms. Most importantly I intend to do no harm to my patients, and will do my absolute best to ensure that intention through. I wish to be able to help my patients the same way that my greatest hero in medicine, my grandfather, did and hope to one day be a doctor of his caliber or better!

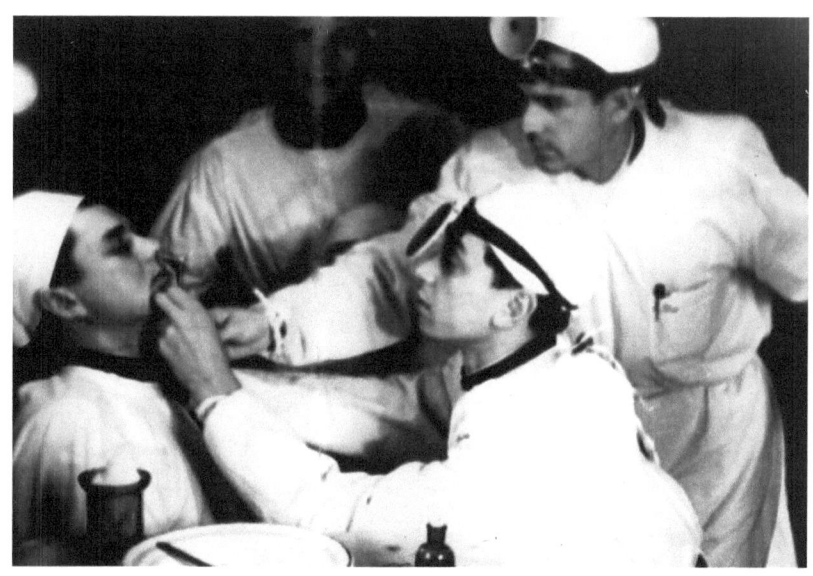

Yuriy Savchenko

Vladimir Ljubimov

Oath of Practice as a Physician

I swear by the Lord above, deities and natural forces in which others find solace and healing, and the most cherished people in my own life, that I will abide by the following Oath in my practice as a physician:

I promise....

> *To always consider the patient as a whole person and never as a set of conditions or symptoms. To learn her name, to remember the important details of her life and home, and to remind myself of the privilege to be allowed her trust. To remember that each face in a waiting or an examination room has a story, and that I must consider and respect that story at all times.*

> *To show empathy and abandon any preconceived judgments of a person's background or life choices.*

> *To keep a patient's secrets and confidence.*

> *To respect and cherish those who educate me ~ whether patients, colleagues, or mentors. To never tread upon any physician, nurse, or person dedicated to the healthcare profession while caring for patients or furthering my career. To be in solidarity with my fellow healers, and to demonstrate that solidarity with my actions.*

Once able, to impart my knowledge, wisdom, and deep regard for others to the next generation of physicians, so that the care of the attacked and the powerless may never step backward according to the whims of politics or public opinion.

To never cease learning.

In daily practice as a physician, I intend...

To do no harm. To respect the patient's wishes, even if they are at odds with my own inclinations, and to know that I did the right thing in doing so. To question those wishes if they are proclaimed by anyone but the patient herself, but to empathize with family members making difficult decisions about their loved ones.

To understand that the definition of harm is not black and white, and to consider each patient individually while adhering to the professional ethics with which I have been trained.

In matters of life and death, to respect the life of the mother over the life of the unborn, unless a mother chooses her unborn over herself.

To steer away from providing medication whenever a condition can be resolved using diet, exercise, and lifestyle changes. To provide patients with "prescriptions" to better their overall health rather than a single symptom, yet to always mind their chief complaints in the structure of their care.

To respect others' health care practices, and to work with traditional or complementary healers as best suited to patient care. To never "cut for stone," consulting with others with greater or different expertise when so necessary.

To stand up for what I believe to be right for a patient or for a group of people, wherever that mission may take me.

And in all things, I commit to hold myself to the highest standards of personal interaction and professional demeanor, to preserve the purity of my life and my arts, and to keep this Oath faithfully.

Amen.

Kristen Roehl

With God as my ever present witness
I commit myself to a lifetime of learning
Doing my absolute best
That my practice of medicine may be
A blessing
To my patients and their families
And a credit
To my colleagues and my profession

Realizing that my motives will be tested
I safeguard my heart
That my intentions may remain pure
And my actions honorable
Acknowledging my imperfections
I welcome the support and help of my associates
Remembering that an ear closed off to advice
Is of detriment to myself and others

With the well-being of my patients
As my unwavering goal and undying desire
I look forward to the experiences to come
Though aware of their challenges
Approaching each patient encounter
With the excitement of my first
Beseeching the Lord to guide my steps
My hands and my heart

A Physician's Daily Prayer

Ashley Prosper

I vow that I will keep this Oath and this contract to the best of my abilities and judgments:

I will listen and learn from those ahead of me in their successes and mistakes.

I will continue to learn throughout my life and career for the benefit of my patients and for my own enjoyment of learning to keep my curiosity strong.

I will teach and lead the students that follow me to the best of my ability to help foster the future of medicine and leave a legacy of positive and intelligent doctors.

I will always be honest with everyone around me by being as complete as possible in all conversations. I will explain everything realistically and thoroughly, but be positive in my delivery and empathetic in my tone.

I will promote prevention as the best form of medicine. I will help motivate patients to make the best decisions for their health. I will not over treat my patients, even when they ask. I will instead fully explain any and all reasoning behind my decision, with the hope that they will see the benefit of choosing better overall health decisions.

I will hold sacred the art of sympathy and understanding, as it is stronger than my intellectual and physical abilities.

I will accept that I cannot be perfect. I will accept short-fallings and be honest about them. I will be comfortable with saying, "I don't know, but I will investigate the research and consult my colleagues to find out as soon as possible and get back to you."

I will help my colleagues when asked. I will offer help when colleagues seem to need an extra hand or guidance in a different field. I will not insert myself or exert my power in specialties other than the one I practice and understand. I will refer those who I do not feel equipped to treat to other colleagues who I feel will do what is best for the patients and their health.

I will talk to my patients in a way that they understand and answer any questions as thoroughly and clearly as possible. I will listen to their problems without judgment of what they divulge, what they do, or what they do not do. I will keep all stories sacred and confidential.

I will do everything I can to maintain a healthy balance between my career, my family, and my social life. I will be a role model of health for my patients physically and emotionally. I will seek my own treatment when necessary and not overlook my own needs.

This Oath will be upheld by me faithfully for as long as I shall practice the art of medicine.

Anonymous

The things I cannot forget when I am a doctor.

I will advocate for all of my patients.

In 1970, no one advocated for this woman.

Women like her, faced many complications during childbirth in Los Angeles. When she had her 4th child in 1976 in Los Angeles, a doctor felt that she should not have any more Mexican children. They placed a diaphragm inside her without her consent or knowledge. She went back to Mexico and had frequent problems, but lacked money to go to the doctor. Eventually, a severe infection landed her in a hospital in Durango, Mexico. They discovered the reason she had not had children in years.

She recovered....somewhat. She had some miscarriages. Then 2 more Mexican daughters. Then a hysterectomy.

I am the fifth daughter.

Remember, your patients are people.

I swear to have relentless faith in my patient's ability to heal.

2012. A 58-year-old mother of 4 was driving to work. She had to pull over because of a headache, nausea, blurred vision and sweating. She decompensated rapidly. Loss of consciousness…seizure. Ambulance to a community hospital. Helicopter to neurosurgery hospital service.

I received a phone call and ran. She went into surgery and one of the two aneurysms burst while the surgeon tried to clip it. Stroke. Second stroke. Craniotomy. Transferred to the Neurosurgery ICU.

I told her son to accept that she might be wheelchair bound. I told him that aneurysms were often the end of lives. One ended my mother's life.

She spent 9 months in hospitals. She walks now.

Remember, expect the worst but always hope for the best.

I swear to be kind to the people that our society has shunned.

Homelessness is often 1 paycheck away.

Remember, groceries on credit cards.

I will look beyond the superficial to identify the roots of the greater problem.

Gangs provide children the family, love, respect, and community that we as a society have failed to provide for them.

Remember that individuals who seem so different from you have the same desires, hopes, and dreams as you.

I commit to educating my patients in vocabulary they understand.

It is our duty to explain, educate, and communicate in a language, volume, and register our patients find appropriate.

Remember the home visits to the little girl with eosinophilic esophagitis and bed bug bites. They did not have money because the father had been deported. The family was hospitalized for 3 weeks and received great printed resources in Spanish. No one noticed that mom could not read. Remain passionate about adequate language services.

I will be mindful of doubts, fears, and concerns about medicine and research.

Many people of color are afraid of doctors, either because of lack of experience, education, or because doctors have done something horrible to them or their community.

Remember to be patient with those that are hesitant or scared of medical interventions. Do not belittle their fears no matter how many cases you have seen.

I promise to create safe spaces for those who are afraid.

"Regardless of whether it is physical, emotional or takes some other form, abuse often follows an escalating pattern in which the aggravating behaviors worsen over time. Any behaviors may be used to control or exert power over a partner, and they may be part of a larger cycle of violence and reconciliation." Women Against Abuse

Remember the young girl who came into the ER. Everyone ignored her because she was just another stupid girl who stayed with an abusive guy. "Why doesn't she just leave him?" I overheard in the nursing station. I asked if she was in pain.... I bothered the attending until he gave her medication. She opened up to me. She had left him. He stomped on her face because she did not have money to give him for a high. She required multiple surgeries to fix her broken jaw and shattered hand.

I intend to serve the displaced – poor people, immigrants, refugees, and migrants.

My family immigrated to the US in 1988.

Poor people share an experience that crosses geographical boundaries.

Remember to always find ways to serve immigrant and poor communities.

I promise to interact with all my patients with respect and advocate for their dignity.

Our elders deserve respect. Their hard work got us here.

Remember to be compassionate especially with elderly patients.

I will only use my craft to improve the lives of those people who seek my help.

As long as I hold these promises close to my heart, I will have the honor of being a physician.

María de Fátima Reyes

I swear to fulfill, to the best of my ability and judgment, this covenant:

I swear to practice both the science as well as the art of medicine in all of my endeavors.
I swear to uphold and cherish the sacred patient-doctor relationship, keeping the words I hear and the things I see at the bedside private and to myself unless as mandated by law.
I will not see disease, but a human being.
I vow to honor and respect the teachings bestowed upon me by my medical educators, be they clinicians, researchers, patients, and peers.
I will endeavor to give hope and to endeavor to the best of my academic and humanistic abilities to provide great care for my patients.
I will endeavor to prevent cross-cultural misunderstanding by being worldly in my academic and personal pursuits.
I will not let borders limit my clinical and research pursuits, but rather uphold values that resonate across all cultures, all backgrounds, and all walks of life.
I promise to be ethical in my efforts to fight disease and treat my patients.
I promise to be grateful for my interactions, conversations, experiences with all those that I meet.
I will remember to be compassionate, understanding and patient.
I promise to be true and sincere to myself and others.
I vow to be earnest, but humble in all of my pursuits.
I vow to turn weaknesses into strengths and failures into victories.
I promise to reach for the stars while keeping my feet on the ground.
I promise to speak the language of medicine and the language of those I meet along my journey, be it through words, gestures, pictures, or a quiet understanding.
I will strive to maintain a strong ethic in my profession and personal life.
I promise to move forward and upward in my pursuit of academic breadth and depth, personal growth and development, and interactions with all those who walk this earth with me.
I promise to establish and maintain a close professional and personal network of peers, mentors, friends, and family. I will be well, do good work, and keep in touch.
I vow to be punctual, responsible, and diligent.
I vow to be familiar with the current state of affairs on a national and international scale.
I promise to be a supportive, caring, trustworthy, and responsible son, friend, student, and physician. I commit to and will forever preserve all of medicine's and humanity's code of ethics and honor.

Henry Wu

Every person deserves my full care and attention.
Every patient deserves the most advanced medical care.
Every day I will find something uplifting about what I do.
Every person has an important and worthwhile story.

There is no boundary to what I will do to help my patients.
There is no boundary to what I will do to help my patients finance their medical care.
There is no boundary to what I will do to help my patients educate themselves.
There is no boundary to who should receive medical care.

My convenience comes second to helping someone else.
My convenience comes second to my patient's comfort.
My convenience comes second to helping a team of doctors.
My convenience comes second to fighting for change.

I will continue to educate myself for as long as I am practicing.

I will continue to be involved with the community as a part of my career.

I will continue to reflect and accept criticism as I progress in my career.

I will continue to remember and remind myself of my optimism at this time in my life.

Barbara Rubino

親愛的主耶穌

因為你的恩典 我能在這裡 在你面前受呼召這使命

不枉費你的賞賜 盡全心全力 你的兒女是配得最好的

求你賜我智慧 讓我更深了解 人體的奧秘和復活的奇蹟

求你讓我謙卑 天天在學習 虛心領受生命的道理

讓我可以成為

你的口 你的手

可以擁有你的心

讓我的耳能聽見 眼能看見

施行你的旨意

讓我一生一世 獻上自己 只為你所用

一步一步 跟隨你腳印

因為你的憐憫 我得以經歷 醫治的大能和神國的降臨

與哀哭的人同哭 為受苦的人捨己 永遠不忘記愛的命令

成為你的器皿 宣揚你的公義 服侍你們中一個最小的

被擄的得釋放 瞎眼得看見 叫那受壓制的得自由

Song written and recorded in Chinese by Samuel Phang

My Prayer

Because of Your grace, I can stand here and accept my calling
To not use this gift in vain, I'll do my best, because Your children deserve the very best
I ask for wisdom, so I can understand, the wonder of the body to heal itself
I ask for humility, that I would be teachable, humbly learning about the miracle of life

Because of Your mercy, I get to experience the power of healing
Mourn with those who mourn, give myself for the suffering, and never forget the commandment of love
Becoming Your vessel, carry out justice, serving the least of these
Prisoners are set free, the blind can see, and the oppressed is released

Let me be Your mouth, Your hands, let me have Your heart
Let that my ears can hear, eyes can see, carry out Your will
I give my whole life for You
Step by step following your footsteps

Samuel Phang

I intend to live with integrity and dignity, acting toward others with kindness and honesty.

I vow to return the respect and compassion shown to me by others – not only upon the giver, but forward onto unsuspecting recipients.

I vow to live without fear, anger, guilt or greed.

In doing this, I hope to inspire others to live to their highest standards.

I will allow those around me to be themselves without becoming the object of my judgment. I will validate the world that exists within each individual, unique in both its landscape and its history. After all, true beauty is present in our complexities, irregularities, and imperfections.

I will recognize that guardedness, resistance, and the other difficult behaviors exhibited by others may stem from their wounds. I will be patient, and will be receptive to their story when they are ready to tell it. When they do, I will consider it a privilege to hear it.

I will treasure their tale, locking it within the vault of my mind, never assuming that I myself have the right to give it away, for it does not belong to me. I will respect the privacy of others.

If given, I will accept the opportunity to intervene on behalf of one's health with gratitude.

I aim to heal.

I seek to find fulfillment and learning in all my interactions with others, expressing gratitude toward them, as they are all my teachers, whether they may be professors of science and medicine, moral compasses that help navigate the gray seas of ethics, or those who teach me about the complexities of being human, simply by living and feeling.

I commit myself to a lifelong pursuit of wisdom, integrating the discoveries I make through study with insights gained through experience. I aim to better myself progressively over the course of my life.

I will exhibit loyalty and good will toward my colleagues, my profession, and my institution. I will remember all those upon whom my actions reflect before making decisions - we are never just ourselves. I aim to carry our good name past the many obstacles along our path, through storms and fires to come, up to the highest of heights.

Tanya Jain Gupta

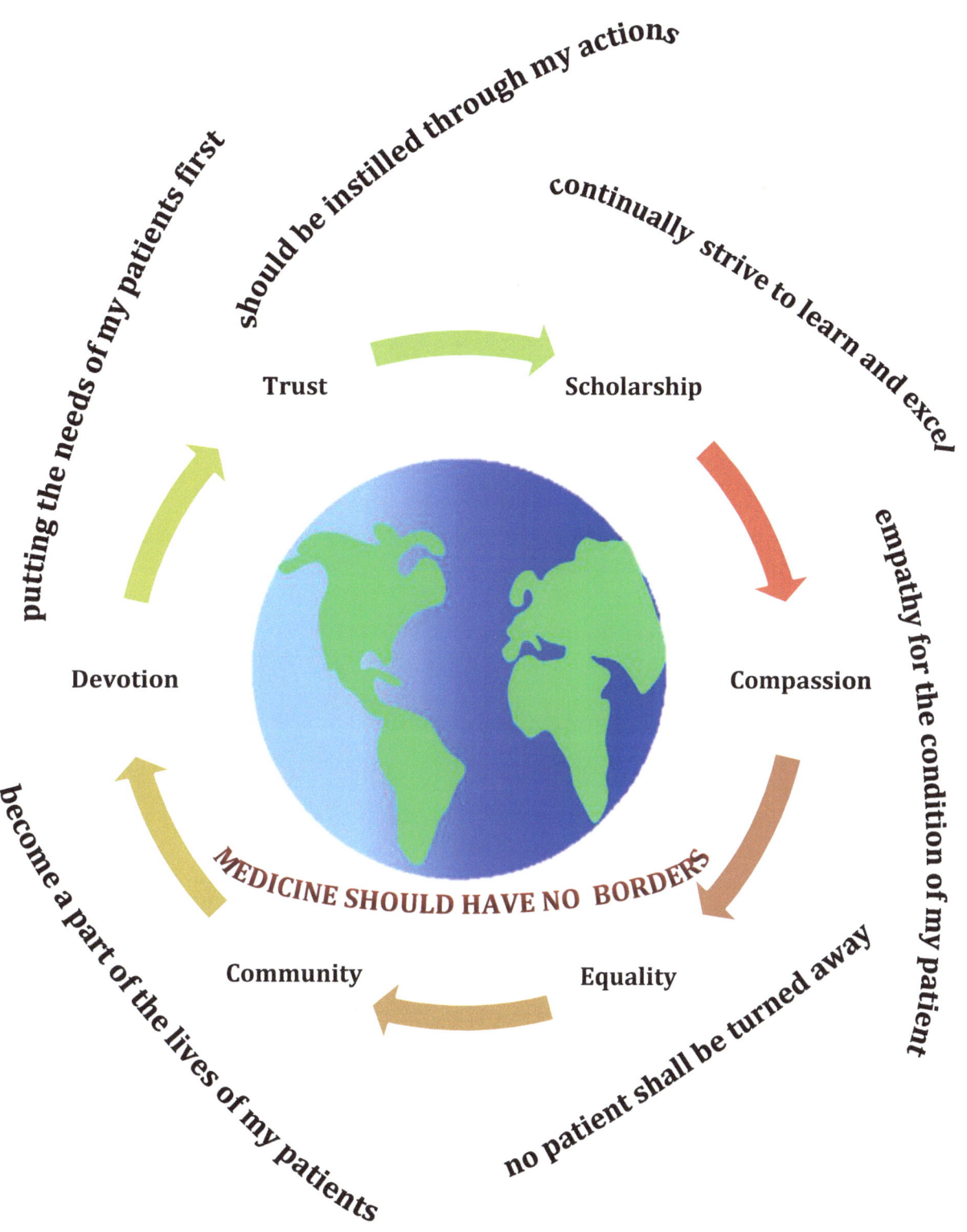

Anonymous

I recognize that the rigors of the path ahead will necessarily change me. I write this oath with the hope that through the peaks and troughs ahead, I have a code that will guide me.

As I assume this humbling responsibility of student of medicine and future physician, I hold myself accountable to the following:

> I vow to remain a lifelong student of the art and science of medicine for the ultimate benefit of my future patients and colleagues;
>
> I commit to assuming duties beyond my role as clinician to make a lasting impact on the many as well as the individual, whether that may be through medical education, scientific investigation, and/or community service;
>
> I promise to remember and respect my teachers in this journey—my professors, my colleagues, and most importantly, my patients;
>
> I trust that I will challenge myself to reflect and grow through both setbacks and successes;
>
> I vow to maintain an unflinching demeanor of compassion, kindness, and patience as I work alongside and for others;
>
> I commit to caring for myself as an individual beyond my identity as healer;
>
> I promise to carry myself with dignity and integrity in my roles as caregiver, teacher, advocate, and scientist;
>
> And I trust that I will continue to challenge my perspectives of those I meet and recognize my role as student of humanism, with the aim of becoming an ever more dependable physician and person.

Anonymous

As a future physician I henceforth decree
To uphold this oath to my best ability
Through medical school, the USMLE,
Residency, and the long hours daily,
I vow to answer my patient's call
Whether it's a concern, a need, or just something small

I will treat my patients with decency and respect
No matter what race, gender, or group they may represent
During their life and after their death,
Their medical needs I will never neglect.

I will support my patients in sickness and despair
By providing treatment and advice with professional flair
And even though diseases may not be curable
I will do my best to make the pain endurable

Patient-doctor confidentiality is of utmost concern
It must be protected so that others won't learn
This sacred bond between the healer and the ill
Is the cornerstone of medicine and it's a duty that I will fulfill

As my years go by, I aim to become a better doctor.
To learn from physicians and patients so that all would prosper
And while the hours may be long and the training may be grueling
There is no other profession that I would rather be doing.

Cato Chan

10 TENETS

1. I dedicate my whole being to the field of medicine to promote the health and well-being of others.

2. I swear to serve my patients and my community with utmost compassion and sincerity.

3. I promise to embrace each day and see each individual with a smile.

4. I hope to always have the patience and open mind to listen and learn from others' unique experiences and perspectives, while respecting my patients' privacy and sociocultural backgrounds.

5. I will remember that patients are human beings, not diseases or medical charts/EMRs.

6. I vow to stay true to my family and the values I have been taught.

7. I wish to conduct my work and live my life with integrity, nourishing myself both physically and mentally.

8. I want my work to further the causes of females, minorities, and all of humanity.

9. I give my efforts to and share my knowledge and experiences with society.

10. I will strive to the fullest degree to overcome barriers within myself as well as adversity in my surroundings to apply not only the science, but also the art of medicine.

Stephanie Chau

I swear by the beauty that leaps forth from every pocket of life and from under every fold of death, and by the love that binds us to one another and weaves our threads into the fabric of the universe, and by the hope that keeps us moving forward when we are tempted towards the abyss of despair, and by all else that I hold most sacred, that I will dedicate myself wholly to the care of people.

I seek to make every act I perform and thought I create an honest attempt to discover who I am and what I can learn to best see people in their entire being, to understand what they suffer from, and to do everything in my power to relieve their suffering.

I let go of any pride, insecurity, doubt, and fear that may hold me back in this journey. I begin with an open mind and an open heart and a desire to meet every challenge with the excitement of learning, rather than the fear of failing.

I believe that every life is precious and beautiful and deserves love, and that what we observe is only one of a set of infinite possibilities. I will learn to see these potentials and guide them to the best of my abilities towards greater well-being, health and happiness.

Now if I keep this oath and break it not, may I enjoy the peace of knowing that I have created meaning in my life by cherishing the lives of all.

Anonymous

From this day forward, I am no longer who I once was. For as many years as I have lived, I have simply lived for myself. By these words alone, I shed my past and enter a new world.

In this new world, I am committed to the well being of others. Whether my fondest friend or mortal enemy, family member or complete stranger, I will treat everyone equally and to the best of my ability.

In this new world, I no longer belong solely to myself. I am a valuable resource to my community and to the world. My skills are the skills of healing and must always be used as such. I will always be honest and forthright to those I serve, maintaining their secrets as if they were my own.

In this new world, I will be the steady, helping hand. Be it sickness, grief, pain, tragedy, strife, or anger, I am committed to alleviating the suffering. I will help even the most downtrodden; it is for them especially that I hone my skills.

In this new world, I will confront the starkest of life's realities. I will know injury, illness, and death. I will have to deliver news that can devastate entire families and communities. I cannot falter in this duty, as I must be the one who represents both life and death.

In this new world, I will be strong for those who have no strength left. No longer shall I quit when I am tired. I must persevere, even in the face of overwhelming odds, to be the beacon of light in the darkness.

In this new world, I am a physician.

Justin Seltzer

I swear by Apollo the Physician, and as God as my witness that I will respect and honor the profession of medicine, and strive to represent myself, and my field, with integrity, commitment and altruism. I will continuously strive to be the best doctor I can be regardless of extenuating circumstances, and aim to devote my fullest attention and abilities to my patients and their families.

I will work collaboratively with my fellow physicians, technicians, nurses, and whoever is working for the betterment of the sick. I will respect these colleagues and teammates, and treat them as allies, throughout the course of my patients' treatments, on and off the clinical battlefield. I pledge to never belittle a fellow healthcare professional, but respect and carefully consider their input into my patients' treatments. I will value their assistance and show gratitude for the learning and experiences that come from working with a diverse group of individuals.

I promise to practice the virtues of integrity and honor, not only clinically, but in my everyday life as a representative of the profession for those outside of the medical field. I will strive to be a living example of moral and ethical action for my family, friends, coworkers, patients, and community. I vow that any person, regardless of race, ethnicity, background, social status, sexuality, profession, or education, I will approach and interact with, only with the intention of doing good. I will treat all patients equally, regardless of which way their actions tip the scale of good and evil, and remain blinded to the character of my patients when it comes to doing my best to heal them.

With the strength of God I will deny any temptation to break my intentions, my promises, my oath. I want to live these promises out, because I know if I do, I am truly being the best physician that I can be, and living up to standards I have set for myself many years ago. I will only use my art to cure, to better, and to aid with any means in my power the health and wellness of my patients. I will feverishly reject opportunities to abuse my position, even if for monetary benefit and furthermore, I pledge to work, not motivated by financial gain, but for the good of first and foremost my patients, and for my family and friends as well.

I will respect my patients, and their privacy, and abstain from disclosing any personal information I may be privy to. I vow to only use their information for the good of their health and wellness. I will aim to make the most of my time with patients, pay the fullest attention to their lives, and give 100% each day, regardless of whatever else I am dealing with in my life. I will strive to create strong patient-physician relationships and be the person that can turn someone's day around.

Ultimately, I want spend everyday with people, working to promote what the World Health Organization so eloquently describes as "a state of complete physical, mental and social well-being and not merely the absence of disease or infirmity." I want to have a longitudinal relationship with my patients that involves not only the treatment of a disease or a condition, but a lifetime journey towards bodily health, self worth, and social stability. I want to empower people to take over their own health, educate the next generation, and be a resource to help patients overcome any barriers to their wellness that they cannot conquer alone. I will work through even the most challenging of situations and know that if I can change even one person's life in a positive way, I have served my purpose on this earth.

Anonymous

ACKNOWLEDGMENTS

The editors recognize and thank all of the student authors who have submitted their personal oaths to this anthology. We hope that we have honored their dedication with this publication.

The instructors of the Professionalism in the Practice of Medicine course at the Keck School of Medicine of USC have generously mentored first year students and encouraged them to offer their oaths to this work.

We acknowledge the support of Pamela Schaff, M.D., Donna Elliott, M.D., and Johanna Shapiro, Ph.D. who believed in this project and encouraged us to publish the collection.

Brian Dolan, Ph.D., our liaison at the University of California Medical Humanities Press, graciously and efficiently made all of the arrangements to produce the anthology. He was thorough in addressing all of our questions, prompt in responding and resourceful in helping us achieve our goal of publishing these oaths. We could not have completed this work without his input and expertise.

This anthology was made possible with a grant from the Arnold P. Gold Foundation. In addition, a grant from the USC Levan Institute for Humanities and Ethics supported this project. We appreciate their confidence in us and our shared interest in medical humanities.

APPENDIX 1

Hippocratic Oath[1]

I swear by Apollo the physician, and Asclepius, and Hygieia and Panacea and all the gods and goddesses as my witnesses, that, according to my ability and judgment, I will keep this Oath and this contract:

To hold him who taught me this art equally dear to me as my parents, to be a partner in life with him, and to fulfill his needs when required; to look upon his offspring as equals to my own siblings, and to teach them this art, if they shall wish to learn it, without fee or contract; and that by the set rules, lectures, and every other mode of instruction, I will impart a knowledge of the art to my own sons, and those of my teachers, and to students bound by this contract and having sworn this Oath to the law of medicine, but to no others.

I will use those dietary regimens which will benefit my patients according to my greatest ability and judgment, and I will do no harm or injustice to them.

I will not give a lethal drug to anyone if I am asked, nor will I advise such a plan; and similarly I will not give a woman a pessary to cause an abortion.

In purity and according to divine law will I carry out my life and my art.

I will not use the knife, even upon those suffering from stones, but I will leave this to those who are trained in this craft.

Into whatever homes I go, I will enter them for the benefit of the sick, avoiding any voluntary act of impropriety or corruption, including the seduction of women or men, whether they are free men or slaves.

Whatever I see or hear in the lives of my patients, whether in connection with my professional practice or not, which ought not to be spoken of outside, I will keep secret, as considering all such things to be private.

So long as I maintain this Oath faithfully and without corruption, may it be granted to me to partake of life fully and the practice of my art, gaining the respect of all men for all time. However, should I transgress this Oath and violate it, may the opposite be my fate.

APPENDIX 2

The Oath of Maimonides[2]

The eternal providence has appointed me to watch over the life and health of Thy creatures. May the love for my art actuate me at all time; may neither avarice nor miserliness, nor thirst for glory or for a great reputation engage my mind; for the enemies of truth and philanthropy could easily deceive me and make me forgetful of my lofty aim of doing good to Thy children.

May I never see in the patient anything but a fellow creature in pain.

Grant me the strength, time and opportunity always to correct what I have acquired, always to extend its domain; for knowledge is immense and the spirit of man can extend indefinitely to enrich itself daily with new requirements.

Today he can discover his errors of yesterday and tomorrow he can obtain a new light on what he thinks himself sure of today. Oh, God, Thou has appointed me to watch over the life and death of Thy creatures; here am I ready for my vocation and now I turn unto my calling.

APPENDIX 3

Adapted Hippocratic oath recited at the Keck School of Medicine of USC's White Coat Ceremony[12]

I swear by Apollo the physician, and all the gods and goddesses, and by whatsoever I hold most sacred, that I will be loyal to the profession of medicine, and just and generous to its members; that I will lead my life and practice my art, in uprightness and honor; that into whatsoever house I shall enter, it shall be for the good of the sick, to the utmost of my power, holding myself far aloof from wrong, from corruption, from the tempting of others to vice; that I will exercise my art solely for the cure of my patients, and will give no drug, perform no operation for a criminal purpose even if solicited, far less suggest it; that whatsoever I shall see or hear of the lives of men and women which is not fitting to be spoken, I will keep inviolably secret.

Now if I keep this oath and break it not, may I enjoy honor, in my life and art, for all time.

References

1. *Hippocratic Oath* Translated by Michael North, National Institutes of Health, National Library of Medicine, 2002. www.nim.nih.gov/hmd/greek/greek_oath.html Accessed March 25, 2015.
2. Oath of Maimonides, translation by Harry Friedenwald. Johns Hopkins Sheridan Libraries. *Bulletin of the Johns Hopkins Hospital*. 1917;28:260-261.
3. Judeus I. Guide for the Physicians-Sefer Muser Harjoin. *Bull Hist Med*. 1944;15: 180-188.
4. Sun Si Miao Zhu. *Qian Jin Yi Fang*. Shanxi Science and Technology Press, 1991.
5. Gellhorn A. Medical Ethics – So What's the Story? *In Vitro*. 1977;13(10):588-594.
6. Dhammika S. *Vejjavatapada: The Buddhist Physician's Vow*. BDMS, Singapore, 2013.
7. Reich WT (ed.) *Encyclopedia of Bioethics*, revised edition, Vol 5. Simon & Schuster, MacMillan, New York, 1995.
8. Oath of the Muslim Physician. *Convent Bull Islamic Med Assoc*. October 1977; 27-30.
9. Hippocratic Aphorism. *Bulletin of the Johns Hopkins Hospital*. June 1913;24:170.
10. Goodgold J. Ethics, Education and Empire. *Arch Phys Med Rehab*. 1997;60:1-3.
11. Orr RD, Pang N, Pellegrino ED, Siegler M. Use of the Hippocratic oath: A review of twentieth century practices and a context analysis of oaths administered in medical schools in the US and Canada in 1993. *J Clin Ethics*. Winter 1997;8:377-388.
12. Keck School of Medicine of USC Adapted Hippocratic Oath.
13. Graham D. Revisiting Hippocrates. *JAMA*. 2000;284(22):2841-2842.
14. Swick HM. Toward a Normative Definition of Medical Professionalism. *Academic Medicine*. 2000;75:612-616.
15. American Medical Association *Principles of Medical Ethics*. Revised 2001. http://www.ama-assn.org/go/ethics-principles. Accessed January 2, 2015.
16. LCME Standards. March 2014 *Standards for Accreditation of Medical Education Programs Leading to the M.D. Degree*. http://www.lcme.org/publications.htm. Accessed March 25, 2015.
17. ACGME Standards *Common Program Requirements*. http://www.acgme.org/acgmeweb/Portals/0/PFAssets/ProgramRequirements/CPRs2013.pdf. Accessed March 25, 2015.

www.ingramcontent.com/pod-product-compliance
Lightning Source LLC
Chambersburg PA
CBHW040905020526

44114CB00037B/67